# PROOFING ROBOTS INVENTION INFLUENCES

## HUMAN BEHAVIORAL CHANGE

JOHN LOK

# Contents

# Preface

Introduction

Nowadays, robotic had been applies to different aspect nu humans, they may include any hospitals' surgeon rooms medical surgeon equipment aspect, hotel room food delivery or hotel front line customer service aspect, shopping center, leisure places cleaning task aspect, student educational service aspect, warehouse logistic goods transport delivery tasks aspect , even general restaurants kitchen cooking tasks aspect, lawyer and accountant firms general bookkeeping, law writing draft clerical taks aspects etc.

How social change influences human behavioral change ? Why human behavior may be influenced by social change? Our individual behavior whether can be influenced to bring negative or positive attitude by social change? I shall attempt to indicate cases to explain whether our individual behavior can be influenced to changed by social environment change. Readers can have more understand how and why social change may influence our behavior in possible. Behavioral economy is one useful and fun social subject. Behavioral economists ususally research how and why human behaviors may influence economy growth or recession, or how and why economy environment changing factor may influence human behavior changes.

Can robotic invention help employers to improve efficiencies or raise productivities in order to raise economic growth or it can bereplaced low-skillful workers to cause any low-skillful jobs are done by artificial robots to achieve many low skillful workers will lose jobs and unemployment ratio will be influenced to raise in our societies?

I write this book aims to discuss whether future robotic invention may help our societies improve economic growth or it can bring recession. This book includes two parts to discuss, this part indicates what reasons to explain why robotics may influence our societies economic growth as well as second part indicates what reasons to explain why robotics may bring our social recession both in possible. Readers can make individual analysis to judge whether robotics can influence our social economic growth or it may bring recession more.

In my this book, I shall attempt to explain how and why ecommerce may be one kind network human job. Also, I shall indicate reasons to explain why human network behavior may bring direct or indirect influences to economy growth or recession in our global societies in macro and micro economy view. Why leisure changing environment may influence human behavior , even economic environment changes. I shall indicate cases to explain any possible human social activities may bring direct or indirect influences to cause our social economic growth or recession in consequency in possible. I hope that my readers can feel more understanding whether what real meaning of behavioral economy is the relationship between our behaviors and our economy.

# Prologue

Contents

# ONE

# HOW ROBOTS CHANGE HUMAN WORKING BEHAVIOR

Why can (AI) be applied to help businesses to predict consumer behaviors ? How can artificial intelligent tools predict consumer behavior in vehicle market retail industry ?What is (AI) consumer behavioral prediction tool? How any why will (AI) tool assist manufactures to attempt to predict consumer behavior before and after consumption occurrence? First, I shall indicate how to apply (AI) tool to predict vehicle product consumer behavior case example.

Nowadays, many vehicle manufacturers hope their vehicles can attract to vehicle buyers to choose to buy their vehicles. However, there are many different brands of vehicles to provide to them to choose, so the vehicle market competition is very serious.

How to judge their different kinds of vehicle price which is reasonable acceptance to attract vehicle buyers to choose to buy the brand of vehicle manufacturers' any kinds of vehicles, e.g. fast speed sport style vehicles, comfortable and slow speed common cars, for four passengers common small size or more than four passengers common large car size? How to evaluate the vehicle prices issue is important factor to influence vehicle buyers' choices. Either if the brand of vehicle price is too high to compare brands, it will influence many vehicle buyers choose to buy other brands' vehicles or if the brand of vehicle price is too low, it will influence vehicle buyers feel this brand's vehicle's quality is worse to compare to other vehicle brands' similar vehicle products.

Thus, if the brand of vehicle manufacturers can predict how to design vehicles which can attract many vehicle buyers to choose to buy whose any vehicle products. What are future vehicle buyers‘ favorable vehicle styles? Then, the vehicle manufacturer can concentrate on manufacturing the kind style of vehicle products to sell already. It will reduce its vehicle manufacturing investment risk.

How to apply (AI) tools to predict vehicle buyers' behavioral consumption model? Whether artificial intelligent tools can predict automotive buyers‘ behavioral consumption model and predict future trend. In fact, automotive brands and dealerships are facing an increasingly competition when attempting to manually gathering the vast quantities of data required to create customer focused programs that increase retention, ultimately new sales and service automotive business. Building a based on that client's intrinsic needs and interests to any kinds of automotive vehicles at any given time. This is especially true in the automotive industry where the time span between purchases is measured in years. Because vehicle buyers would not like often to change their old vehicle to another new one. So, their decisions to buying another new vehicle, the time is usually after one year, even longer time. Hence, it seems any vehicles won't be frequent consumption products to the owned at least one vehicle family consumers (vehicle buyers).

Hence, how to predict vehicle consumers' taste or preferable which styles of vehicle choices issues is very important. If the vehicle manufacturers can not manufacture any attractive vehicles to sell easily in this year. Then, it will lose time, money in this year because it won't know when the owned least one vehicle users or non-owned any vehicle users who will decide to buy one new vehicle or change another new vehicle ensure. The different brand vehicle dealers will possible wait more than one year to attract them to buy their vehicles if their styles are not attractive to compare other brands of vehicle competitors.

However, artificial intelligence and machine learning can help any vehicle manufacturers to find solution to solve patterns in highly to solve patterns in highly complex data-sets that are beyond the capability of a human brain, and then building and automatically acting on the customer insights it generates.

Given the automotive customer need for individualized communications, this technology is positioned to become a critical component of any successful vehicle retailer's domestic or/and overseas vehicle markets. How can vehicle manufacturers and retailers use (AI) to enhance their vehicle marketing campaigns? How will (AI) affect their vehicle sale marketing strategy? What criteria would they use when selecting on (AI) solution?

Vehicle consumers today are able to quickly access different brands of vehicle information, research vehicle products and reviews, negotiate prices and compare one vehicle brand or retailer to another resulting of the brands of vehicle customers. At the same time, the rise of " big -data mining", wearable devices that track user's every move and preference and greater contextualization in advertising and social media has resulted in consumer expectations of individualized. Thus, it seems that (AI) tools can be used to gather " big-data" and then they can make human's mind to analyze how to design kinds of vehicles to satisfy vehicle buyers‘ needs.

As automotive vehicle marketers can apply (AI) tools to achieve messaging strategies to meet the needs of this new generation of informed vehicle consumers, using data from a variety of sources to move from a variety of sources to move from mass- messaging to more personalized messages aimed at particular vehicle buyer segments, e.g. fast speed sport vehicle buyer segment, slow speed comfortable small size or large size of buyer segment. However, when 90% of vehicle marketers believe having a single vehicle buyer view is important, only 6% have achieved it.

However, one of the main issues vehicle marketers facing is the lack of capacity to efficiently sift through and analyze the massive vehicle buyer amounts of data required to create vehicle buyer individualized vehicle customer experiences easily. This is especially difficult for automotive dealers, the long periods between purchase cycles, and the highly considered nature of the vehicle purchase means that each vehicle dealer needs to not only track a large number of potential vehicle customers for an extremely long period of time, but each of those vehicle customers will generate a huge amount of different kinds of vehicle behavioral consumption data as they research their next vehicle purchase. However, by choosing the right (AI) technological tools and programs , vehicle dealers can solve this big data gathering challenge into a major advantage.

For Forrester vehicle brand example, vehicle consumers have more power over the Forrester vehicle brand's reputation than ever before. Mayne, L. (2014) indicated that Forrester calls this new (AI) tools is the " age of the vehicle customer", a 20 year business cycle in which the most successful vehicle enterprises will reinvent themselves to systematically understand

and serve increasingly powerful vehicle consumers. To win in this new age, Forrester declares companies must become vehicle customer obsessed and the only sustainable competitive advantage is knowledge and engagement with customers, such as (AI) gathering data knowledge.

Thus, the biggest challenge vehicle businesses currently face is not the collection of a large quantity of vehicle consumer data, but what to do with that data once they have it. Even at a large vehicle data research firm, the data sets are often too big for a single analyze, or even a team of analysts to sort through and draw conclusion from. However, enter artificial intelligence and machine learning , an efficient technology solution that can continuously find patterns in highly complex data sets that are way beyond the capacity of a human brain and then automatic drive action based on the customer insights is generated.

What is (AI) machine learning tool? Machine learning is a type of (AI) that learns from data and is not explicitly program. Think Amazon, face book. Machine learning serves up relevant content based on an individual vehicle purchase behavior and experiences. More simply, machine learning is a computer program that can learn relationships between data, subject those learnings to errors functions, and then learn from its errors. The program in effect, trains itself.

Lee, T. (2016) explained that "Thus, (AI) tools can learn deep a more advanced branch of machine learning inspired by how our brain's nervous function, has also been found to be especial effective in identifying patterns from data."

When this way sound is complicated from a vehicle dealer perspective, the implementation of a marketing program driven by artificial intelligence can take care of these tasks in an automatic vehicle fashion with little to no manual intervention required from the staff at time vehicle stores.

In practice at a vehicle dealership, the program will continue track vehicle customer behavior online, merging that data with any offline source ( like CRM or DMS data) and then analyze this aggregated vehicle buyer data set to predict what vehicle customer may be shopping for and what information they might like to relevance from different kinds style of vehicle design photos.

Artificial intelligence refers to complex in vehicle market, machine learning that posses the same characteristics of human intelligence and that have all our sense, all our reason and think just like human do. Besides, machine learning is the practice of using algorithms to collect and examine data,

learn from it, and then make a determination or prediction about something in the world.
The machine is " trained" using large amounts of data and algorithms that give it the ability to learn how to automatically perform a task with increasing accuracy. Otherwise, deep learning is primarily based on artificial neural networks inspired by our understanding of the biology of human's brains.
Deep learning breaks down tasks in ways that enables machines to assist us with increasingly complex tasks, driverless cars, better preventive healthcare and more accurate product recommendation ( including vehicle recommendations). So, such as why (AI) technology can be applied to predict how vehicle consumer behavior changes to bring to judge whether vehicle consumer will like what kinds of vehicle styles next year. Then, vehicle manufacturers can gather overall vehicle consumer data to analyze and conclude the more accurate vehicle design direction for next year any new design vehicle manufacturing products.
Thus, (AI) machine learning can help vehicle manufacturers to solve how to design any new vehicle products challenge. A vehicle is both one of the most important and carefully considered purchases the majority of people will ever make in their lifetime. It is also a purchase that tends to be fundamentally tied to a person's identify and view of themselves. As the same time, vehicle consumers changing lifestyles result in changing vehicle needs, e.g. the young sport car enthusiast matures into the family driver.
Automotive dealers need to remember that vehicle customers and prospects are individual human beings with risk, complex and ever-changing lives factors, these factors will influence every vehicle consumer why who feels has vehicle purchase need, and how who choose to buy the first vehicle if who decided to buy the first vehicle.
The (AI) technological customer behavioral prediction tool seems to be the best vehicle salespeople in the world are those that know every one of their vehicle customers. Their likes and dislikes which style of vehicle design, preferences and changing tastes to vehicle choices. The capacity of the human brain, however, limits us from achieving this type of vehicle sales and frequent turnover at vehicle dealerships often results in the further loss of vehicle salespeople along with their vehicle customer relationships and knowledge. In this competitive vehicle environment, machine learning enables platforms to assist the vehicle sales team by tracking the vehicle consumer behaviors of each vehicle customer, learning and memorizing

their preferences and predicting their future vehicle purchase needs.
Finally, I recommend that for a vehicle dealerships marketing platform to make their customer engagement efficient and fully-functional, I should be able to: applying (AI) tools to track every vehicle customer behavior across the web, connecting to a society of data sources, CRM, DMS, third-party, web vehicle brands, social email, click etc., aggregating and accurately cross-reference data from a variety of sources, leveraging this data to drive insights on a mass scale, as well as on an individualized basis, driving actions and automatically direct customer engagement via multiple channels based on where each customer is in their individual lifecycle.
How can (AI) provide businesses with better-informed decisions?
I shall explain how (AI) technology can provide businesses with better-informed decisions to drive top-line growth, deliver meaningful experience for customers and smooth their path along the consumer journey. The widely understood definition of (AI) involves the ability of machines or computers to learn human thinking, reasoning and decision-making abilities.
A Narrative science study in 2015 year identified that (AI) was being used primarily in voice recognition, machine learning virtual assistants and decision support. This study also highlighted the many branches of (AI) and that techniques and their definition are used interchangeably. It is possible that (AI) can be used to gather big data , then to analyze to help businesses to predict consumer behaviors. For example, one of the most common techniques is machine learning, where algorithms are used to perform tasks by learning from historical data. Another growth branch of (AI) is natural language procession.
However, during 2017 year, search engines will begin to factor additional behavioral data into prediction of customer behavioral results, such as the user's history of searches and locations and previously captures conservations. Artificial intelligence will use this information to power predictive search results, e.g. predictive future consumer's choice behavioral processing for any kinds of businesses.
Predictive search will improve the quality of search results, and provide new insights into consumers' behavior and the moments which matter to them. Search will give recommendation into tailored how consumer individual choice in consumption process. Several of the largest online platforms already use machine learning to improve predictive consumer behavioral search results.

For example, Google's rank brain technology adds research by understanding the context in which the consumer has entered it. Over time, rank brain will learn further from user behaviors Amazon's DSSTNE ( pronouned destiny) learns from shoppers' purchasing habits and consumption behavior to offer better product recommend actions, which Amazon can offer before a consumer has entered anything into the search bar. However, this technology is not independent of human input. For example, Google engineers will periodically retain the rank brain system to improve the models it uses. For another example, in 2016 year , Apple computer revamped its photos app to allow consumers to search for specific items in the phots, they want to find, not just dates and locations. Each photo that an intelligent phone or intelligent pad user takes goes through 11 billion computations, so that photos can understand exactly what is the photography.

It seems that in future, (AI) machine learning will allow search to evolve even further. Search engineers will deliver refined recommendations to their business users and use less human input to predict consumers' needs. For IBM computer example, it indicated 90% of the data that exists today has been created in the last two years. This huge explosion of data gives brands the opportunity to quickly spot and react to the latest trends, fashion and fads among its clients and potential clients. This will allow companies to better engage with younger consumers, who gain influence access to the latest trends, and use the brands. They associate with to help define who they are as individuals. Thus, brands have to identify and make use of them before consumers move on, but the vast quantity of data available makes. This a resource-intensive task. For next example, Lesara, a based online clothes store, uses this machine learning to inform its product decision often gathering information from internal and external sources. When its trends -spotting shoes. Lesara has a range of over 20 styles and sells hundreds of pairs a day. It focus on giving consumers, the very latest trends allow Lesara to develop on average of 50,000 new items each year. It compared to 11,000 old items each year. Thus, (AI) brain seems to human brain to own analytical ability to predict consumer behaviors.

For another example, Lesara is one online clothes store, uses machine learning decisions after gathering information from internal and external sources. One of its most popular products, shoes with LED started life when its trend spotting software flagged up a blogger wearing similar shoes. Now Lesara has a range of over 20 styles and sells hundreds of pairs a day. Its focus on giving consumers the very latest trends allows Lesara to develop an average of 50,000 new items each year, compared to 11,000 for its competitor Lara. it seems (AI) machine learning can help Lesara business to predict what kinds of shoes design or style that shoe consumers will prefer choose to buy in future shoe market trend. Thus, Lesara can predict shoe consumers' taste successfully and it can manufacture many attractive style of shoes. (AI) machine learning can gather global past shoe consumer's shoe shopping experiences, then analyzes to make conclusion to give lesara recommendation successfully. This will make the experience more enjoyable for shoe consumers and allow Lesara to advert whose different new style or design of shoes to deliver them move relevant messages by understanding the context of the experience.

However, (AI) machine learning will have this risk who manufacturers need to concern if they applied this technology to predict consumer behavior. It is on sample consumers' privacy issue, in order to avoid complaint chance occurrence. However, machine learning can tie this data together to identify which f the billions of devices are being used by individual consumers. This helps brands understand how consumer engagement and actions can be attributed to different messages in different contexts and at different time. So, machine learning can help brands to build confidence to promote their products by any advertisement channels. When, this new (AI) machine learning technology can conclude how to design their products to be the most attractive, due to it has more accurate to predict consumer behaviors to compare human themselves prediction judgement effort. It seems that (AI) machine judgement effort is more accurate to compare to human judgment effort.

For example, google is moving away from cookies and using logged in data to track and make to users. It plans to expand the scope of the brand lift tool from online video. Thus, consumers are responded will to shippable context, finding it persuasive and easy to navigate by (AI) machine learning decision. For example, fashion brands can aggregate their You tub videos and blogs into a mobile context marketing experience, such as brand centric context into a personal shopping activity gives the shopper an experience, who are likely to remember and tell their friends about any new style of products design promotion from these internet advertisement channels after (AI) machine learning tools' styles of product design recommendation.

What is (AI) deep learning techniques to help farmers to raise food growth
The (AI) deep-learning technology leads to performance enhancement and generalization of artificial intelligent technology. It influences the global leader in the field of information technology has declared its intention to utilize the deep-learning technology to solve environmental problems, such as climate change. So, it will help agriculture farming businesses can raise any plant food: vegetable, fruit, rice which grow up very easily if farmers can apply (AI) deep-learning technology to solve environment problems to influence their plant food grow. If the whole year seasonal change is very good and it is suitable for any plant food to grow in farming land easily, e.g. rain is enough and soil is enough for any plant food to grow in the farm lands. Then, fruit, rice, vegetable etc. agriculture businesses will have much beneficial attribution to global farmers.
The question is how to use deep-learning technologies in the environmental field to predict the status of pro-environmental consumption. We predicted the pro-environmental consumption index based on Google search query data, using a recurrent neural network ( RNN model). To certify the accuracy of the index, we compared the prediction accuracy of the RNN model with that of the ordinary least square and artificial necessary network models. For example, the RNN model predicts the pro-environmental consumption index better than any other model. we expect the RNN model to perform still better in a big data environment because the deep-learning technologies would be increasingly as the volume of data grows. So, deep-learning technologies could be useful in environmental forecasting to prevent damage caused by climate change to influence any rice, vegetable, tomato, potato, fruit etc. different plant food grow in any countries' farming land easily.
For South Korea example, over 800 government agencies spent 2.2 trillion Korea won on eco-products in 2014 year. However, green products are rarely purchased outside these agencies. This phenomenon occurs because there is a gap between consumer attitudes and behavior , that is environmental attitude is a major factor in decision making vis-a-vis the consumption of " green" food and services ( Jorea Ministry of Environment, 2015). Therefore, it is necessary to understand those consumer attitude, that will lead to sustainability-conductive behavior and consumption.

Environmental consumption prediction

Recently, many researchers have studied pro-environmental consumption and household indexes as well as suicide rate predictions using messages posted by internet users on Google trend, Tweets etc. channel. Whether can environmental consumption be predicted by (AI) deep-learning technological internet channel? How can impact the pro-environmental consumption attitudes of green policies? Korea scientists estimated pro-environmental attitudes using search query data provided by Google trend and confirmed through regression analysis, that pro-environmental attitude has a positive correlation with the pro-environmental attitude index. They also explained that environment-friendly attitude of residents plan an important role in policy making. In the past, most household consumption indexed were calculated through surveys, but (AI) deep-learning technological tool " big data" have recently gained research attention ( Lee et al. 2016).

It seems that (AI) deep-learning technology can help agricultural export countries‘ farmers , e.g. US, UK, Canada, New Zealand, Australia, Japan, China, India etc. they can predict environmental behavioral consumption to any rice, tomato, potato , fruit, vegetable etc. plant food consumers. The beneficial advantages to them include as below:

(a) Assuming they know their countries' weather, when it has less rain to cause drought or when it has more rain in any seasonal time in the year. They can choose not to grow any kinds of above these plant food to avoid loss.

(b) They can make any kinds of above these plant food price raising after their prediction of these bad seasonal time to cause their plant food shortage supply challenge. Because these plant food consumers‘ demand number is more, but the supply of these above plant food supply number is less. However, due to they had predicted when the bad seasonal time can not allow them to grow these above plant food before. So, they have enough time to grow many these above plant food number in predictive good seasonal time to prepare to supply to their plant food import countries' plant food consumers to eat. Thus, these predictive environmental consumption plant food export countries can raise their plant food price to sell to them. When, the other non-pre-predictive environmental consumption plant food export countries can not supply any one of those plant food to them to eat, due to the bad climate to cause them can't grow any one of these plant food to export to sell.

Thus, (AI) deep-learning technology can be applied to predict how to raise the plant food supply number in order to raise price to the import plant food countries consumers to eat, due to they feel difficult to buy these plant food to eat in the bad climate seasonal time in whole year.

(c) (AI) deep-learning technology can help climate scientists to find what reasons cause their countries; rain sudden increases or cause their countries‘ rain sudden decreases. After its gathering data analysis, it can assist climate scientists to find solution methods to attempt to control the rain level can be right falling down level to let agricultural export farmers who can grow their plant food to sell to agricultural import countries in whole year.

(d) The agricultural export countries’ farmers can apply (AI) deep-learning technology to help them to choose whether growing which kinds of plant food in that whether climate time to earn more plant food consumption number more easily.

Due to the agricultural countries climate will often change, for example, tomato, potato, rice, fruit etc. plant food can be adapt to grow in more rain time, but vegetable can not be adapt to grow in more rain time. If farmers can apply this technology to predict when it will have move rain or when it will have less rain to fall down in their countries. Then, they can choose to grow which kinds of plant food number more, in the suitable seasonal climate time in order to raise plant food growing number productivities to supply to sell to satisfy any agricultural food import countries‘ demand effectively.

(e) (AI) deep-learning technology can help agricultural import countries to solve agricultural food shortage challenge in long term. When this technology can be popular to base applied by the agricultural plant food export countries. It will solve global agricultural food shortage challenge. For example, when one agricultural export countries’ farmers can popular accept to apply this technology to predict when to grow which kinds of plant food more to rise number productivities to sell. e.g. vegetable, fruit, rice Besides another agricultural export countries‘ farmers can also accept to apply this technology to predict when to grow plant food, e.g. potato, tomato to raise number productivities to sell. Then, they can concentrate on growing the specific kinds of plant food in order to raise the specific plant food number productivities in every seasonal change time every month. Then, global agricultural plant food supply must be raised, due to these predictive environmental change farmers can know who ought grow which kinds of plant food to sell to raise number productivities.

How can apply (AI) technology to help book shops to decide book sale price

(AI) digital channel can be applied to help businesses to evaluate whether how much the product price is the most attractive to persuade consumers feel it is the most reasonable price to sell. It helps consumers to feel which brands of products which ought change the price to let consumers to choose to buy the brand of product. It can be applied to predict whether how many consumer numbers can be increased or decreased when the brand of product's price is variable. It aims to give opinions to help any brand of product manufacturers or sellers to judge whether which price is the most reasonable to let consumers to accept to choose to buy the brand of product in popular.

Thus, (AI) price measurement technology can be preference to be applied online communication ecommerce and mobile phone internet platform aspect. As businesses can enter their past products prices data and past customer number data into computer or mobile. Then, (AI) price measurement technology can gather these data to analyze these product prices and past customer number to compare their prices variable changing range level to find their price variable difference to measure to make conclusion about every product's price variable changing will influence how many customer number increase or decrease changing to choose to sell their different kinds of products more accurate. Then, (AI) price measurement software will help them to analyze all past price variable changing data to compare whether which price range can let customers to feel it is more reasonable and attractive to influence them to choose to buy the product among different brands of product choice.

Because any product's price is one important factor to influence consumers to choose to buy the product, instead of quality, durability, shape, appearance, color, brand familiarity etc. factors. Any online businesses with a focus on Asia should considerate (AI) customer care, and virtual shopping experience, whereas is Europe and North America still value face-to-face and/or real human interaction over (AI) or virtual worlds.

For example, Amazon publish has applied (AI) price measurement technology to help authors to decide how much every different topic of e-book or paper book price, it can attract the largest number of readers to buy. Any one author only needs to type whose book name to Amazon publish author himself/herself Amazon website. Amazon publish (AI) price measurement learning machine will help them to auto-calculate and judge how much e-book or paper book price is the most attractive and the most reasonable in order to increase reader number to buy their e-books or paper books to read. So, (AI) online price measurement machine will gather past similar book names and past every similar book readers' reading times and the number of readers to give opinions to let every author to judge whether his/her very new e-book or paper book ought charge how much price to the e-book or paper book which can attract many readers to choose to buy. Although, it is not ensure that the e-book or paper book price must let readers to feel it is the most reasonable price to choose to buy in reader's view point. However, it has other factors to influence readers' choice to buy the e-book or paper book, e.g. whether the book content is attractive to public, the author's familiarity, the book's page is enough or not to satisfy readers to read etc. factors. But, instead of all these extra factors to influence readers to choose to buy the book to read. (AI) price measurement learning machine can real give opinions to every author to let them to judge the e-book or paper book different price range whether is too high to influence readers to choose to buy to read or tool low to influence readers feel it is possible poor content book to compare other similar content books. Thus, (AI) price measurement machine can help authors to predict every reader's reading behaviors or reading experience and reading habit from online channel in short time easily. The author only enter the book name to let Amazon publish price measurement machine to check, it will follow past reader's reading habit and reading experience to judge whether the similar all book topic sale record to judge how much price is the reasonable price to attract many readers to buy the book.

Hence, (AI) can be applied to digital channel to help businesses to predict consumer behavior in the future. In the future, mobile/smartphone, laptop, desktop will be most frequent used ecommerce channels to develop online business. So, (AI) can be also applied to these platforms to gather data to make analysis to help businesses to predict consumer purchase behaviors popularly. Due to , ecommerce is popular to global, so digital online and instore channels can be one good channel to let (AI) learning machine to make platform to gather past every online consumer purchase ( buying) experience data to help businesses to build brand personality and having a responsible, positive impact on society.

To apply (AI) learning machine technology to understand customer online purchase behavior, it will raise business e-commerce successful chance: For example, (AI) learning machine can help businesses to gather data to analyze to determine whether short-term or long-term signals in the online consumer behavior that indicate higher purchase intents to let every online business to know. (AI) learning machine can find that online users with long-term purchasing intent tend to save and click through on more content.

However, as online users approach the time of purchase their activity becomes more topically focused and actions shift from saves to searches from online consumption channel. Then, (AI) learning machine will further find that the brand product purchase signals in online behavior can exist weakness before an online purchase is made and can also be traced across different online purchase categories. Finally, (AI) learning machine synthesize these insights in predictive models of online user purchasing intent to the brand of product. Taken together, it's work identifies a set of general principles and signals that can be used to model online user purchasing intent across many online content discovery applications. Thus, (AI) learning machine can help online businesses to gather any online users' click online behaviors data to judge whether there are how many online users will choose to find their online business websites to make final decisions to buy their products from online channels. Then, it will give opinions to help the online businesses to let it to judge whether what are the important website factors will help its online business to attract many online consumers, e.g. designing unattractive website issue, online unattractive product photos issue, unclear website color issue, unclear website advertisement message, contents and words impressions issue, lacking image movement frequent attractive seeing issue etc. different website factors. Thus, online digital channel will be one good choice to apply (AI) learning machine to help businesses to predict consumer behaviors.

Can apply artificial intelligent learning machine " big data" gathering method to predict manufacturers‘ behavioral performance ?

In consumer view point, can they apply (AI) learning machine to predict manufacturers' behavioral performance to judge whether whose products are value to buy. Nowadays, (AI) and big data are reshaping the risk in consumer privacy. For example, consumers want to hide their willingness to pay just as firms want to hide their real marginal cost, and buyers have less favorable information, say a low credit shore, prefer to withhold it just as sellers want to conceal poor product quality. So, it implies that it is possible (AI) learning machine can help customers to gather any manufacturers‘ past sale performance, e.g. how many complaints or appreciation from clients, product quality etc. sale data to let consumers to make judgement whether it is value to buy to compare other competitors. So, it has risk to the poor product quality of manufacturers. Otherwise, it has benefits to the good product quality of manufacturers. It also implies all manufacturers' privacy is not protected or secret when (AI) learning machine is popular to be used to predict manufacturers' behaviors by consumers.

Information economists suggest that both buyers and sells have an incentive to hide or reveal private information, and these incentives are crucial for market efficiency. Data technology that reveals consumers type could facilitate a better match between product and consumer type, and data technology that helps buyers to assess product quality could encourage high quality production.

Thus, (AI) big data technology can also assist consumers to gather different manufacturers' data to compare what their advantages and disadvantages of their products are. Then, consumers can make comparison to choose which brand of product is the suitable to whom to buy in these more choice consumption market. (AI) learning machine will gather similar brand their products' data to analyze to make conclusion to let consumers know or feel to make final judge to find what advantages or disadvantages of these sample brands of similar products' comparison from internet. On the other hand, it means that manufacturers can gather consumers' past purchase behaviors or purchase experience from (AI) big data gathering method to record and analyze to give opinions to let manufacturers to know what reasons or factors influence consumers choose not to buy their products from internet.

(AI) big data gathering consumer behavior prediction method can give these benefits to manufacturers and consumers both, such as: New concerns arise because (AI) technological advance which have enables reducing cost of collecting, storing, processing and using data in mass quantities extend information beyond a single transaction. These advances are often summarized by the big data, it means charge volume of transaction-level data that could identify individual consumers by itself or in combination with the datasets.

The popular (AI) takes big data as in input in order to understand, predict and influence consumer behavior. Modern (AI) is used by legitimate companies, could improve management efficiency motivate innovations and better match demand and supply. But (AI) in the wrong hand, also allows the mass production of fraud and deception. Since , data can be stored, traded and used long after the transaction. Future data use is likely to grow with data processing technology, such as (AI) big data gathering consumer and manufacturer behavioral prediction method from internet channel.

Thus, future (AI) big data learning machine can also help consumers to choose the best brand of manufacturer's products among different brands of manufacturers products choice to compare their past sale performance from internet. They can apply (AI) big data statistic method to gather all different manufacturers' similar products past sale data to compare their advantages and disadvantages to make the best decision to choose to buy which brand of product is the most suitable to them to buy to use. It seems (AI) big data can also help consumers to predict any manufacturers' manufacturing behaviors or manufacturing performance whether they are improving their product quality or are deteriorating their product quality. Thus, (AI) big data tool is also important to help customers to predict future the different brands of manufacturer performance will have improvement in possible.

Thus, I believe that artificial intelligent "big data" gathering method can be suggested to be applied to attempt to predict consumer behavioral changes in global business environment, the reasons are as below:

On the consumer's beneficial hand, Consumers can apply this method to attempt to gather any global manufacturers data to be analyzed by this artificial intelligent learning system. Then, it analyzed all the different brands of specific similar product manufacturer' data to compare what are the range of the best past manufacturing history and sale data to the group of best manufacturers, and what are the range of the better past manufacturing history and sale data, and what are the range of the good past manufacturing history and sale data, and what are the range of the common past manufacturing history and sale data. Finally, the (AI) learning system will compare all the specific similar product, e.g. mobile phone or computer, television, car etc. different kinds of specific products of global manufacturers to conclude the result is such as whether which brands will be the best manufacturers to let the consumer to buy the television or mobile phone or computer or car etc. different kinds of products. It can make more accurate judgement to compare general human's phone or questionnaire surveys investigation method, newspapers, television, radios, internet searches etc. different manufacturing news or data gathering channels to find which brands are the most worth confidence to consumers to choose to buy the specific product in the global consumption market.

On the manufacturers' beneficial hand, manufacturers can apply (AI) data gathering method to predict consumer emotion and buying behavioral changes more accurate. For example, the vehicle manufacturer, it plans to gather data to predict potential driving fast speed sport vehicle consumers' preferences trends in order to make the accurate judgement how to design its sport vehicles to attract many sport vehicle buyers who will choose to buy it's brand of any driving fast speed sport vehicles. It can attempt to apply (AI) intelligent learning system to gather global different brands of sport vehicle data concerns that all past driving fast speed sport vehicle buyer's preference of sport vehicle design. Then, the (AI) intelligent learning system gather global different brands of driving fast speed sport vehicle which had ever been purchased by the different country's driving fast speed sport vehicles consumers. After, it can compare divide the range of similar driving fast speed sport vehicle design and similar price to be different groups. The (AI) intelligent learning system can attempt to follow the past number of different brands of driving fast speed sport vehicle buyers to calculate how many driving fast speed sport vehicle buyers who choose to buy the brand of driving fast speed sport vehicle as well as it will analyze and make judgement to find whether the cheaper price reason attracts the different countries sport vehicle buyers choose to buy the brand of driving fast speed sport vehicle or the attractive design reason attracts the different countries sport vehicle buyers choose to buy the brand of sport vehicle or fast speed reason attracts the sport vehicle buyers choose to buy the brand of sport vehicle.

For example, although some brands of driving fast speed sport vehicle manufacturers' prices are very high, but they can still attract global many sport vehicle consumers to buy. Whether all sport vehicle's attractive design is the main factor to influence them to buy or whether it's fast speed is the main factor to influence them to buy or whether it's safe confidence it the main factor to influence them to buy or it's familiarity brand is the main factor to influence them to buy. (AI) intelligent learning system will attempt to make judgement and analysis to conclude whether the attractive design factor is the main factor to influence many sport vehicle consumers to choose to buy the brand of sport vehicles.

Otherwise, for another example, although some brands of driving fast speed sport vehicle manufacturer's prices are low, but they can not still attract many global many sport vehicle consumers to buy. Whether all vehicle's unattractive design is the main factor to influence them choose not to buy their fast speed driving sport vehicles or whether the unsafe factor is the main factor to influence them choose not to buy their fast speed driving sport vehicles or whether unfamiliarity brand is the main factor to influence many consumers choose not to buy their fast speeding sport vehicles.

Thus, when (AI) learning system had helped the fast speed sport vehicles manufacturer to gather all different brands of fast speed driving sport vehicle's past sale data and price data, design of different sport vehicle, e.g. color choice, method of style, comfortable chair styles and chair sizes and what kinds of steel material to manufacture the sport vehicles data and driving safe and accident occurrence data and the data concerns what reasons of the past complaint to brand of sport vehicle manufacturer from its sport vehicle buyers. Then, it can make more conclusion to give more accurate opinions whether which brands of fast speed driving sport vehicle manufacturer(s) whose sport vehicle design is the main factor to attract consumers choose to buy its any driving fast speed sport vehicle products really. Thus, it seems that it can make more accurate judgement to compare television survey, questionnaire survey to gather data concerns how to design the fast speed sport vehicle to attract consumers to choose to buy the sport vehicle manufacturer's planning sport vehicle products. I believe that (AI) learning system can help the sport vehicle manufacturer to make more accurate conclusion or judgement how to design its fast speed driving sport vehicles to attract it's consumers more easily.

How to apply (AI) tool to help businesses to bring consumer positive emotion ?

If (AI) tool can be confirmed to apply to predict consumer behavior, then I can conclude that it can be attempted to apply to predict what the factor(s) of the product itself can cause the consumer has positive or negative emotion, so the manufacturer can attempt to avoid the bad factors cause to bring negative emotion to influence the consumer chooses not to buy the product more easily, such as vehicle product case.

Economists refer to the consumption desirability is as " utility" and the product or service consumption decision making is arose

influenced by maximizing utility only. However, they neglect consumer individual immediate emotion change will also influence the consumer individual consumption decision consequently. Expected emotions are those that are anticipated to occur as a result of the outcomes associated with different possible courses of action. For example, if a potential investor, were deciding whether to purchase a stock, who might imagine the disappointment who would feel if who ought it and it reduced its price. Otherwise, whose emotion would experience , such as regret if it increased in price, but who does not buy it before the stock rise its price. However, I believe nowadays technology, in the future one day, (AI) tool can be attempted to assist consumer psychology profession or marketing research profession to assist them to find what are the bad factors to influence consumers choose not to buy any manufacturers' products. Then, when the manufacturer

can discover what are the bad factor(S) cause(S) consumers who do not choose to buy their products, then the manufacturer can raise whose product of consumption desirability or " utility" to raise whose product's consumption decision making is influenced by maximizing utility. Hence, (AI) tool will be possible to find what the bad factor(S) to cause consumers do not choose to buy the manufacturer's product in order to raise the product's utility to bring consumer positive emotion to choose to buy its product in possible. SO, (AI) tool will be one consumer psychological emotion prediction tool to assist any manufacturers

to help their products to build positive emotion to any consumers in possible.

The key feature of expected emotions is that they are experienced when the outcomes of a decision materialize, but not at the moment of choice, at the moment of choice, they are only feel about future emotion. Such as consumption case, if the consumer chose to buy the product or consume the service before it's price is increased. Then, the consumer will feel happy and it is worth to purchase or consumer the service as well as the consumer's expected emotion is positive before who decides to buy the product or consume the service, because who believes or feels the product or service's price will be raised in short term, e.g. after one month, one week. Thus, it means that if the consumer does not believe or

feel or predict the product or service's price either it will increase or decrease in short term, whose emotion will be negative, those negative emotion will influence who does not decide to buy the product or service, it is possible that who feel it is not worth to buy the product or consume the service immediately. He She will choose to consume the service or buy the product to wait it's price is decreased later. it seems that the consumer's positive or negative emotion will influence who decides to buy the product or consume the service later or earlier. Thus, it has close relationship between the consumer individual immediate purchase or consumption decision and positive emotion or negative emotion ( either expected emotion or immediate emotion influences).

Consequently, if (AI) tool can help any manufacturers

to predict when its product price ought to be increased or decreased in order to attract consumer to choose to buy its product. Then, it can help any manufacturers to build positive expected emotion to attract consumers to choose to buy its product more easily. For example, when the (AI) tool can predict when the consumer expects the product price will fall down, then it can give ideas to the manufacturer to raise up the product price in the month, then it predicts many consumers expect the product price will fall down after six months. So, the product price will not be fall down after six months. So, many consumers will feel disappointment and they will choose to buy the product if the manufacturer

decided to raise the product price after six months. Then, the higher product price will cause many consumers worry about the product price will continue rise up, so they will prefer to choose to buy the product immediately after six months because they afraid the product price will continue to rise up in the year. Then, I assume that (AI) tool has effort to predict when consumers feel the product will rise up or fall down, then it can give ideas to the manufacturer when to rise up or fall down the product price in order to attract or persuade many consumers choose to buy the product in different period in the year.

Psychologists indicate that immediate emotions, by contrast, are experienced at the moment of choice and fall into one of two categories. Integral emotion, like expected emotions, arise from thinking about the consequences of one's decision, but " integral emotion", unlike expected emotions are experienced at the moment of choice. Such as purchase stock case, the share buyer might experience immediate fear at the thought of the stock's losing value. " Incidental emotions" are also experienced at the moment of choice, such as a consumer predicts the product or service price whether it will be risen up or fallen down. If he/she feels the product or service price will fall down after next month and he/she will choose to buy the product or consume the service. But consequently, after next month, the product or service's price won't fall down absolutely.

Then, he/she will have incidental emotion to influence whom to choose whether he/she ought buy the product or consume the service, due to the product or service price is not still fall down. Otherwise, he/she is fear the product or service will not fall down in short term. Even, it will increase price later. Hence, whose incidental emotion will have possible to influence whom to choose to buy the product or consume the service after one month, if the product or service's price is still not increased absolutely. So, (AI) tool can be attempted to apply to predict when the product price ought need to be raised or fallen down in order to attract consumers to choose to buy the manufacturers' product in different period.

Economists indicate utility an individual consumption with an outcome might arise from a prediction of emotion: For example, a dinner eater might choose a higher utility to an Italian restaurant dinner than a French restaurant dinner because who anticipates being happier at the former, even the former's dinner price is higher than the French restaurant.

So, such as this restaurant dinner case, if one (AI) tool can assist the French restaurant owner to find what factor(S) cause(S) the dinner consumers do not choose to go to its restaurant to eat its food, e.g. high price factor, bad taste factor, bad wait service performance factor, bad cooker's cooking skill factor, poor advertisement promotion factor, poor familiar factor, poor location or poor eating environment etc. different factors. Then, the French restaurant owner can find methods to avoid the bad factor(S) cause(S) many dinner consumers do not choose to go to whose French restaurant to eat dinner more easily.

The question is that whether the positive emotion factor can influence the consumer changes whose mind to choose to consume the more expensive service or buy the more expensive product. To answer this question. it depends on whether the consumer has an imperfect understanding of whose own tastes or the consumer has a perfect understanding of whose own tastes to the product or the service.

It means the consumer will choose to buy the product or consume the service, even it's price is higher than other general similar products or services if who has a perfect understanding of whose own tastes to the product or service. Otherwise, who won't choose to buy the product or consume the service, due to it's price is higher than other general similar products or services if who has an imperfect understanding of whose own tastes to the product or service. So, it seems that the consumer's negative or positive emotion arise will be influenced by whose perfect or imperfect understanding of whose own tastes to the product or service factor.

It concludes that whether how much degree of the consumer's utility to the product or service. It is not the only one important factor to influence the consumer to choose to buy the product or consume the service. Otherwise, the consumer's imperfect or perfect understanding own tastes to the product or service factor will influence the consumer to arise positive or negative emotion to make final purchase or consumption decision immediately. It will be one more consumption influential factor to lead the consumer to make the final consumption decision making immediately. So, future (AI) tool ought to be innovate to own how to judge good taste or bad taste for any food in order to predict food consumers to choose to buy the food manufacturer's any foods more attractively.

(AI) tool technical innovation in cruise tourism
immediate positive emotion influence to
cruise travelling consumers in entertainment
service industry

Can apply (AI) tool to cause positive emotion to cruise tourism consumers? Cruise tourism industry is the most influential emotion industry example to influence cruise travelling consumers' travelling entertainment choice. I shall indicate some evidences how it's innovation will influence cruise travelling consumers' emotion to be changed to positive from negative immediately as well as to prove how the cruise traveler higher utility feeling to the cruise tourism provider is not the main factor to influence whom to choose the cruise provider to consume whose cruise journey service arrangement.

Nowadays, cruising has become one of the fastest growing sectors within tourism, cruise service providers need have themselves unique different entertainment service arrangement to satisfy every cruise travelling consumer individual needs in order to attract every one to choose whose cruise arrangement easily, e.g. meals, activities, entertainment and varied destinations create one-stop holiday shop, reasonable competitive ticket fare. Hence, it seems it is one exciting emotion industry. If the cruise service provider can bring positive emotion to influence many cruise travelling consumers immediately. The, even it change higher service fare to compare other similar cruise service providers. I believe it won't influence them to choose other similar cruise service providers if it can often bring immediate positive emotion to its cruise clients during they are staying in its cruises or during they have left its cruises, but they will often remember or won't forget to enjoy their cruise service provider's happing time forever. Hence, if (AI) tool can be attempted to help cruise entertainment providers to arrange different cruise journeys for varied destinations , to arrange different entertainment facilities, to arrange the different taste food to satisfy different countries age cruise consumers' needs. Then, the (AI) tool will assist the cruise providers to bring positive emotion to let every different countries age cruise consumers to feel satisfactory in order to choose to the cruise providers' cruise entertainment service more attractively.

4.1 How can apply (AI) tool to predict cruise service providers bring positive emotion to their clients?

Future, (AI) tool can help any cruise providers to design these kinds of any one entertainment service arrangement to satisfy the cruise provider's customers' needs.

There are different special interests cruising , such as wellness at sea, freighter cruises, river cruises. It has increased the attractiveness of cruising: Romance is for lover cruise traveler target, luxury is for rich cruise traveler target, exotica is for enjoyment exciting feeling traveler target. So, every kind of cruise traveler target will have different kind of cruise entertainment service to satisfy their needs. If the cruise service provider can provide the right and attractive cruise entertainment service to satisfy the specific cruise target. Then, it will bring the positive emotion to the specific cruise target consumers more easily.

Cruise travel was shaped for mass tourism. Prices have been very differently segmented. There are basically four types of markets ( Biederman, 2008):

- Contemporary market: On board fun and amenities are playing important role and destinations have secondary importance.
- Premium market: This category is more expensive than the contemporary category and where the destination has same importance as on board amenities.
- Luxury market: It was once dominant type of cruise tourism, but now it has only a small portion of the industry. Generally, it is the most expensive cruise category and usually it takes longer than average cruise days.
- Adventure/exploration: It refers relatively long cruises with special and exotic places where the destination is the main purpose of the trip.
- European cruise travel: Duration takes more five days than worth American travel duration. There is a tendency on European market during the years that duration of travel is getting shorter. This short demand of is explained with the strong demand of customers ( Hensen, 2003). Beside this, it is most likely that cruise companies try to convince tourists with short

haul travels instead of long term cruise trips for more expenditure.

Thus, I believe that even, the cruise service provider charges higher ticket which won't influence cruise consumers who do not choose its entertainment service on its cruises. If it can arrange the attractive cruise entertainment facilities and destination journey arrangement, staying days arrangement to satisfy different specific cruise target market needs absolutely in order to bring whose emotion to be positive to it's service provision. Then, the cruise service provider will attract many potential cruise clients to choose its cruise service absolutely. Otherwise, if it only bring negative emotion to its cruise clients, it will not attract many potential cruise clients to choose it or loses its old cruise clients, even, its cruise ticket price is needed to decreased in order to raise competitive effort.

In conclusion, I believe that future (AI) tools need to learn how to bring cruise consumers to arise individual immediate or expected positive emotion, this positive emotion consideration is more important to compare to how to reduce cruise ticket price in order to attract cruise clients in global cruise competitive cruise industry.

4.2 Differentiation through the characteristics of cruising route method from (AI) tool route judgement

Future, (AI) tool can attempt to help any cruise entertainment service providers to judge how to design different route to attract different countries age cruise clients' choices to satisfy their cruise journey entertainment needs. The determinants of the cruising route's characteristics ( functional, social, and emotion) is important factor to influence the cruise service provider's success. Cruising product is no longer selected primarily for the cruising service, but for the content of cruising route. So, the cruising route will influence the cruise consumer individual emotion, because it is the main service need for every cruise consumer.

The approach called the " land sea cruising in product development" is increasingly becoming an area of interest, e.g. determining the direction of the effects of the individual cruising route characteristics on service value's perception , and providing an evaluation model of the route's perception , and indicating significance variables of attraction.

The questions that cruise planners need to know: How does each of the identified determinants affect the overall perceived value of the cruise route? How the overall perceived value of the cruise route affects customer behavior intentions?

Because different routes factor will influence cruise consumer individual emotion changing seriously. It means the ship has become only a tool, when the offered route whose attractiveness highly influences the impression of the guests has become crucial.

Consumer behavior in cruising segment includes all the activities and influences in the selection of the specific cruise route. There activities result in decisions and actions related to a defined price, selection and reselection of cruising company ( Cannot, Brink and Brijball, 2006).

4.3 How to apply (AI) tool to arrange cruise route planning have close relationship to influence cruise consumer emotion?

Firstly, use value of cruising routes is based on the subjective experience, and shows how individuals assess the route during, or immediately after sailing. It is affiliated with the benefits that cruising guest realize by choosing a route , and it is subjective because it depends on the individual assessment ( photo taken on the route for one guest presents just a family souvenir, and for professional photographers are embodied financial capital).

Secondly, the utilitarian value is also subjective-oriented and is tied on the point where the inner and us ability of cruising routes are compared with the sacrifice of the client ( money and time). Finally, the value is considered as the outcome of the comparison of scarifies and personal benefits, which is resulted in essentially utilitarian nature.

Hence, route design is the main value of cruising tourism and it is primarily determined and analyzed from the aspect of observed customers. Otherwise, the cruise is only one tool to be caught for the cruise passengers, whether the cruise can let whom to sleep comfortable , providing what kind of food to them to eat, what kind of entertainment facilities are provided to them to play, these issues are not more important to compare how to design route to bring them to travel to anywhere to enjoy in this cruise journey factor. Because how to design the route factor can bring each cruise passenger to influence them to feel either negative or positive emotion directly. The whole route journey planning is the most influential factor to influence the cruise passengers to feel whether they ought choose it's service again or not in the future.

In economic utility or immediate (expected) emotion aspects, whether which is more influential to excite consumption. To analyze whether it is economic utility or immediate ( expected) emotion more influential to excite travelling consumption. It depends on the travelling consumer individual consumption choice is in which situations. For example, if the travelling industry's general travelling consumer individual consumption decision is concentrate on emotion influential aspect, such as cruise entertainment industry, hospital care service industry, theme park entertainment industry, movie watching entertainment industry etc. Above all these industries have same nature, it is service. So, it seems that service industry's main influential factor is immediate ( expected) emotion influence, it is not economic utility influence. Otherwise, product sale industry's main influential factor is utility.

(AI) judges travelling consumer utility factor

Travelling consumers' emotion and utility factor will be seem to child's emotion and utility factor. For this toy choice situation example, parent choose to buy one toy to give whose child to play. They usually considerate which kind of toy is attractive to their child whom like to play. In many different kinds of toys choice, if the child likes to choose the kind of toy to play. After the child's parents had purchased the kind of toy to let whose child to play one period time, e.g. six month. Then, when the child feel that who has need to buy another new toy to play, due to he/she feels bored to play this toy. So, it seems that the child feels this toy has less utility or it's utility is decreased. So, he/she expects whose parent can buy another new kind of toy to let whom to play. It also implies that it is not emotion factor to influence the child to feel boredom and unfunny to play this kind of old toy after six months. It is the product's utility factor which can not attract the child to play it any more. So, this old toy's utility is decreased when this child spends six months to play it. This toy's value is only six month utility to this child to play. Otherwise, if this kind of toy is bought by another parent. It is possible that the another child like to play this kind of toy one year or more. So, it's utility to another child is one year or more period. So, product's utility period is difference, it depends on how long time of the user's satisfactory time.

As this toy case, the child's decision will influence whose parent choose which kind of toy to buy to whom to play. Usually toy price is not difference too much. Parent won't consider when the toy price will be increase or will be decreased to influence their emotion to decide not to buy the toy immediately. So, when the child like to play the kind of toy, even the product's price is more than other kind of toys, and the parent feel it is possible that the kind of toy's price will be fallen down later. They will still choose to buy the kind of toy to let their child to play, they won't be influenced not to buy this kind product by later cheap price factor. So, immediate emotion is not the main factor to influence this parent does not choose buy this toy at this moment. Otherwise, utility factor will influence when the parent will buy another new kind of toy to provide to this child to play. If the child enjoy to play it only three months, after he/she will feel bore and he/she will tell whose parent to buy another new kind of toy to let whom play when the fourth month is beginning. So, it implies that if the kind of toy product can have more attractive utility time, then it can attract many parent to choose to buy it among different kind of toys. Thus, when this kind of toy's utility time is longer time. Then, it is possible that it can influence many parents choose to buy it's different style or design of similar kind of toys to let their children to play. In general, when many parents accept to buy this kind of different style or design of similar toys to give their children to play. Due to it's popular long time utility factor, it will influence children like to play it longer time to compare other kind of toys. Consequently, it will influence parents do not need often spend too much money to buy other kinds of toys to give their children to play. So, longer time utility factor to the product can attract many consumers to choose to buy the kind of product to compare lesser time utility factor to the product. Hence, it proves the explanation why utility factor is the main influential factor to influence the consumer choose to buy the product.

(AI) judgement tool of Medical care and utility case

Travelling consumers' emotion and utility also seems to medical care patient's emotion and utility. Medical care is an input in producing health, it is subject to law of diminishing marginal productivity. Health yields utility to the consumer. It is subject to law of diminishing marginal utility. It bring this question: Does either the patient's emotion or the medical care service or medical care product utility which one can influence the patient's hospital choice more?

To answer this question: We need to know medical care is one kind of nursing care service in hospitals or clinics and medical care product is one kind of medical care product sale from merchants, e.g. medicine or medical equipment. So, in medical industry which has different kind of medical care services to provide to patients in hospitals or clinics as well as which has different kind of medical care products sale, e.g. medicine or wheelchairs, heart health measurement equipment etc. different medical care products in medical health industry.

In medical care aspect, it is one kind of any medical care service to patients from hospitals or clinics. So, medical care is an input in producing health service to patients from hospitals or clinics. When the patient is admitting to hospital or clinic, who needs to see doctor and the doctor need to give the right medicine to the patient to eat to ill whose illness. Even, if the doctor feels the patient whom needs to live hospital for one time period. Then, the hospital nurses must need to take care the patient during he/she is living in the hospital period. Consequently, if the patent can be health in short time, e.g. within one week leaving time, then he/she will be shortened time to leave the hospital in next weak. Otherwise, if the patent can not be health in short time, e.g. within on week, then he/she needs to live the hospital more than one week, even, one month, three months or more. So, the staying hospital time will influence the patent's emotion to feel whether the doctor's effort. If he/she needs to live the hospital long time, he/she will bring negative emotion to feel the doctor's medical effort is not good. The doctor's medical effort can not achieve or satisfy whose expected emotion during whose staying hospital time.

Thus, medical care is an input service in producing health, it is subject to law of diminishing marginal productivity. When the patient does not need to live the hospital longer time, the patient will feel more satisfactory to the hospital's doctor and nurses' care effort as well as the patient can give less money to spend the expenditure to live the hospital. So, the law of diminishing marginal health productivity will explain the hospital will shorten time to the staying days of the hospital to the patent as well as the patient's care expenditure will be decreased when he/she only needs to live to the hospital in short time. Otherwise, the patient needs to live longer time in the hospital, it means that the diminishing marginal health productivity to the hospital, the patient's staying hospital days will be increased and the patient's medical expenditure to the hospital will also be increased. It will bring negative emotion to patient and why this negative emotion factor will influence the patient would choose another hospital to live or find other doctors to see if he/she felt illness in future one day.

In medical care product aspect, health yields utility to the consumer. It is subject to law of diminishing marginal utility. Because patient needs to buy different kind of medical equipment to use or medicine to eat to attempt to cure whose illnesses. So, if the patent can choose the right medicine to eat from the doctor's recommendation or if the patent can choose the right medical equipment to use from the doctor's recommendation. When the patient buy less number of medicine to eat, then he/she can be health or when he/she buy the medical equipment to use, then he/she can be health. Then, he/she will spend less money to buy medicine to eat or medical equipment to use and he/she can be health in short time. Then, the medical consumer will feel the medicine or medical equipment has good utility to satisfy whose medical needs. So, good medicine and good medical equipment can only need short time and less money to let the medical patient to be health.

How robotics can excite cruise entertainment customer's immediate ( expected) emotion

As cruise entertainment case, every cruise journey must provide fixed stay days on the cruise to let every cruise passenger to play to every cruise journey. So, cruise passenger can not change or extend whose fixed stay day choice in every cruise journey, when they had caught the cruise to go to sea on the day. Is implies that cruise entertainment has none longer time utility factor which can influence each cruise passenger's choice to each different design of cruise journey arrangement.

If the cruise passenger feels very satisfactory and enjoyable to the last time of specific cruise journey arrangement, e.g. five days and four nights New Zealand and Australia cruise journey. Due to this cruise journey can bring positive emotion to let the cruise passenger to let whom feel that he/she can not forget or remember this happy cruise journey forever.

It is possible that this time happy five days and four nights New Zealand and Australia cruise journey will bring positive emotion to influence this cruise passenger to choose to find this cruise service provider to help whom to arrange this same cruise journey or another similar cruise journey again after one month, or three month, or six month or one year or more. Due to this cruse passenger felt this cruise service providers' cruise journey design arrangement can satisfy whose needs and it can achieve whose expected emotion to be positive. So, this cruise entertainment industry must be immediate ( expected) emotion influential factor more than time utility factor to influence the cruise consumer's cruise service provider and cruise journey choices.

In conclusion, it is not only utility factor can influence consumption decision. It is emotion factor can also influence consumption decision. It depends on situations whether the consumer is choosing to buy one product or consume one service. If the consumer is choosing to buy one product, how long time of the product's utility factor which will influence the consumer choose to buy which product. If the consumer feels the product can give longer utility time among other similar products, then he/she will have more chance to choose to buy the product. Otherwise, if the consumer feels the product can give lesser utility time among other similar products, then he/she will have less chance to choose to buy the product. If the consumer is choosing to consume one service, emotion factor will influence the consumer choose to find which service provider to consume the same service or similar service. If the service provider can provide excellent service to the consumer, then it will bring positive immediate or expected emotion to whom and it can attract the consumer to choose the service provider again. Otherwise, if the service provider can not provide excellent service to the consumer, then will bring negative immediate or expected emotion to whom and it can not attract the consumer to choose the service provider again.

Why and how (AI) judgement tool can judgement what utility factors are to influence travelling consumer emotion

In emotion and utility both aspects, they include these situations. I shall explain how and why emotion and utility factors can influence travelling consumption behaviors in these different situations as below:

(1) In the first situation is brand factor, the brand image, product quality, product knowledge , attitude and

(2) brand loyalty intangible factor will attract the consumer individual purchase.

For example, luxury

products, e.g. luxury fashion brands of clothing. Brands like Zara from Spain and H&M from Seweden began to produce catwalk-style fashion at low cost offering consumers of luxury fashion alternatives at low prices.

Nowadays, the luxury fashion sector is the fourth largest revenue generator in France, and one of the most remarkable sectors in Italy, Spain , the USA and the potential markets of China, Russia and India. The luxury industry has increased having a huge youth in demand. The luxury consumer have much choice in products, shopping channels and pricing of luxury products. It has possible relationship of age, gender, income and other demographic factors with purchasing intentions to influence the rational and emotional buying behavior regarding luxury fashion products.

(2) The second situation concerns the decision-making of make or female consumers are possible experience an emotional desires and cognitive ( reasoning) mind in purchasing choice process. Their emotion includes negative or positive buying emotion and mood management and cognitive process components include cognitive deliberation, planning buying with the exception regard for the future.

University had been using analysis of variances tests, male and female students were found significantly different with respect
affective process components including positive buying emotion, and mood management and cognitive process components include planning buying.

Significant differences were also found between the following product categories: shirts/sweaters, skirts, coats, underwear, accessories, shoes, electronic hardware, computer software, music , CD or DVDs, sports, memorabilia, health /beauty products and magazines/books for pleasure reading. No differences were found in regard to suits/business wear and entertainments.

The investigate proved that some products will have different emotion influence to cause female or male students whose final consumption decision to buy the kind of product. So, the difference od male and female students will have emotion influence to make purchase decision to buy the product in consuming choice process.

(3) The third situation concerns search advertising factor, e.g. online search to influence consumption behavior. Advertising is possible one method to persuade the consumer to choose to buy the product, even the consumer does not know the product exists. For example, proper cloth, a company based in New York, has a site on the social networking site Facebook.

Whenever the company posts a new photos of its clothes, all its face book " fans" automatically receive the information on their own face book pages. "We want to hear what our customers have to say." It seems online advertisement is a potential promotion method to promote any new attractive products to sell to let publicity to know to buy. Internet is one popular communication tool to be used by youth today. So, when one company can have one website to let any youth to find and enter to the website to discover any new things easily. It will cause many consumption chances to let online potential clients to attempt to choose any products to make purchase decision from online advertising tool easily.

How does the role of advertising influence the purchase decision process? Needs and motivations are the starting points of purchase decisions. In fact, advertising is a communication of photo image, sound image, and word advertisement image channel to persuade consumers to choose to buy the brand of product or consume the brand of service between the merchant and its consumers from television, radio, newspapers, magazine, movie etc. channels.

Why does advertisement influence consumer
choice? For this case example, when one buyer waits until more information is gathered before making a decision. The time, two types of cost are involved. First, there are psychological opportunity costs experienced by consumers who are deprived of the product who need and are consequently in a state of psychological tension.

As time elapses, this psychological tension becomes more frustration. Second, buyers experience costs with the information-gathering efforts. They must invest time and energy to visit several retailers, seek out and read advertisements, or inquire for other opinions about the best product to buy.

These delayed decision costs considerably increase as time elapses. The buyer must seek information until it is felt that a search for additional information will bring about more costs than benefits. So, an advertisement is reaching a potential buyer when who is seeking information will have a greater impact, since the buyer is spending time and effort needed to seek out this information himself and he is less likely to find other competing and advertisements to obtain the additional information.

In general, buyers are generally more responsive to different brand advertisements, when they are seeking information on these brands. This is why the becomes a choice target for the advertiser provided the advertiser can identify and locate them. Thus, a client has interested and is in an information-gathering stage is asked.

Then, the advertiser takes advantage of the consumer's having identified him or herself to send a series of informative and persuasive messages or to send a salesperson who will try to conclude a sale. Thus, advertisement gives a chance to let consumers to gather information to choose the best product to buy or the most excellent service to consume.

What of situations do merchants need advertisements promotion? The short purchase cycle markets are characterized by routine purchase decision processes or by limited problem solving when a new brand is introduced on the market, e.g. coffee, bread , sugar, soft drinks, canned vegetables and household and beauty care products fall into this category. Another irregular purchase cycle markets are characterized by products that are purchased more or less regularly , e.g. cookies, cake mixes, wines, food products. Finally, long or unpredictable purchase cycle markets include all durable products, such as cars, household appliances and furniture ( products from which occasions of purchase can't be predicted, which most consumers buy only occasionally). Hence, these kinds of products ought need advertisement promotion specially, due to advertisement can build brand image to let consumers to know. Especially , it is a new brand of product. In conclusion, advertisement will be one good channel to let new product to introduce its brand to let clients to remember in minds heart.

Consequently, future (AI) tool needs to learn how to design different kinds of advertisement to attract consumer attention to the product, needs to learn how to find what the bad factor(S) which cause(S) many consumers do not choose to buy the product, needs to learn when the product price needs to be raised up or fallen down in order to bring consumers' positive emotion to choose to buy the product immediately. SO, if (AI) tool can be invented to own itself effort to design different kinds of methods to predict consumer behavior in order to bring their positive emotion to the product successfully, then the manufacturer can earn positive consumer emotion advantage from the (AI) tool assistance for long term benefit to its product.

Main barriers influence artificial intelligent travelling consumer behavioral prediction

In future, it is possible that these barriers will influence how to apply (AI technology) to predict travelling consumer behavior in success. The barriers may include: Lacking of a (AI) digital data gathering vision and strategy, lacking of efficient workforce readiness, (AI) technology constraints., non reaching (AI) consumer behavioral prediction mature stage, time and money and resource constraints, law and regulations prohibition to develop (AI) consumer behavioral prediction bug data gather technology.

However, the recommendation of solutions to attack the barriers to influence artificial intelligence consumer behavioral prediction not success, it may include gaining employee buy in to participate and develop (AI)

consumer behavioral prediction technology, making customer experience to a concern (AI) big data gather questionnaire investigation, providing compensation, training to employees in order to achieve (AI) travelling consumer behavioral big data questionnaire investigation research digital technological goals and strategy, task senior leaders manage any (AI) digital big data gather technology changes, putting policies and (AI) big data gather digital technology in place to support a fully remote, flexible workforce in any (AI) digital big data gather questionnaires research projects, teaching all employees how to code/understand (AI) big data gather consumer behavioral prediction software development, appointing a chief (AI) officer to manage any (AI) big data gather customer behavioral prediction projects and automate everything and encourage customers to attempt experience to self-service and (AI) big data gather questionnaire research to earn beneficial consumption aim after they gave feedback to any (AI) digital questionnaire researches. So, in the future, the (AI) digital big data questionnaire researches can include these industries surveyed, such as automat m financial services, public healthcare, private healthcare, technology, telecoms, insurance, life sciences, manufacturing, media and entertainment , oil and gas, retail and consumer products etc.

Hence, in the future, any of these industries can attempt to apply (AI) digital big data gather technology to predict how and why consumer behaviors will change in order to avoid reducing consumer number threat occurrence.

## 3.1
## (AI) digital data gather technology predicts food consumer behavior's main barriers

It seems food consumer' (AI) predictive barriers is similar to travelling consumer. What are the main barriers to food industry? When the food manufacturer applies (AI) big data gather technology to predict food consumer behavior? The barriers include that the food manufacturer / provider needs to decide whether when the right time is applied to the right (AI) digital big data prediction tool channel to find the right food consumers to be chose to full food consumption satisfactory questionnaires, how to gather multi-class food consumption classifiers on real-world food consumers transactional data from the food sale domain consistently to show the critical numbers of different kinds of food items at which the predictive performance most accurate? So, any food manufacturer /

provider's advanced in (AI) digital data gather warehousing and management technologies can provide that opportunities for food business to enhance long term relationship with the food providers' clients.

However, food industry's (AI) digital data gather aims to improve food customer product targeting, increase food customer loyalty and food purchase probability to the food supplier. To effective identify, understand and satisfy the needs of their food customers, the food suppliers need to develop the right (AI) digital questionnaire questions and find the right food customers to fill every right questions from every digital questionnaire at the right time through the right channel.

Above of all these, they will be the barriers when one food supplier expects its (AI) digital data gather questionnaires which can conclude the most accurate prediction concerns any kinds of consumer food product choices. So, such as (AI) digital data prediction model, it is needed to incorporate into the food market segmentation, food customer targeting, and food challenging decisions with the goal of maximizing the total food customer lifetime. For example, (AI) big data gather transaction data is reasonable and accurate for building predictive models. Transaction data can be electronically collected and readily made available for data mining in lot quantity at minimum extra costs.

Suggestion to apply (AI) prototypes of food customer profiles method to predict food customer behavioral changes. Prototypes of food customer profiles mean to be extracted from the discovered bins and multi-class classifies models are built using those prototypes. The learned models can than be used to predict the class of food customer profiles ( e.g. restaurants, school canteens, supermarkets etc. food suppliers) based on their food purchases. The approach is validated on the case study of a food retail and food service company operating in food and beverages market.

So, a food customer profile, it is a description (AI) data gather tool will record every of food customer using available information, which help in understanding their background and food consumption behavior. (AI) data gather tool can well develop every food customer profile, every food customer data is essential in food market analysis as they aid food suppliers in saving time and money by highlighting the real potential food consumers whose needs are to be met rather a range of individuals.

So, (AI) data gather tool can record every food consumer profile and every can be factual or behavioral food consumption. A factual food customer profile consists of a set of characteristics for (AI) big data gather record, e.g. demographic information , such as food customer name, gender, birth date, when a behavioral food customer profile consists of what the food customer is actually doing and is usually derived from (AI) digital transactional data gather record.

So, (AI) big data gather record's every behavioral food consumer profile can be much stronger predictor of the future food supplier consumption choice actions of a food customer. Furthermore, the food supplier's (AI) all past food consumer information that make up demographically based all past food customer profiles are expensive to acquire when the information for the food suppliers' past every food consumer food consumption behaviors. Moreover, food customer profile can be recorded to make real food purchase every time. So, when the food supplier finds the past food consumer's record from (AI) big data gather tool. Then, it can make more accurate judgement whether past every food consumer has chose to buy its food to eat how many times every year in order to predict whether its every past food consumer will choose to buy its foods how many times next year in possible. If the next year, its every past food consumer's consumption time to the food supplier is less than its current year consumption time. Then, the food supplier can attempt to find whether what factors to cause the past food consumers do not choose to increase food purchase times to the food supplier in current year. The factors may be possible be the food supplier's food prices are raised, food quality or taste is poor, the different kinds of food supply is shortage challenge, the food supplier's consumers lose confidence to buy the food supplier's foods to eat, when (AI) big data gather tool can help the food suppliers to find what the main factors to cause the past food consumer number to be reduced in order to predict how future food consumers' behavioral changes will be influenced from the food supplier's competitors in the global food supply market. Hence, (AI) big data gather tool can help every food supplier to attempt to find what the main factors to case the food supplier's food consumer number to be reduced as well as it can help the food supplier to predict how the food supplier's potential ( past not every purchase its any food consumers) food consumers who can be persuaded to choose to buy its foods to eat by learning what the main factors influence.

In conclusion, (AI) big data gather tool can help the food supplier to find what the main factors influence its past food consumers do not choose to buy its food more times or find what the main factors will attract its potential ( not ever buying its foods consumes) food consumers to choose to buy the food supplier's foods to eat.

3.2 The challenges of (AI) big data gather shaping the future of retail for consumer industries

The future of retail for consumer industries' (AI) big data gather challenges are similar to future travelling industry's entertainment consumption challenges. Another challenge of (AI) big data gather is that how to shape the consumer behavior to let business owner to feel or know oe predict. It means that how it express it's conclusion or opinion for every consumer behavior after it had gather all big data in any data gather period, e.g. three months, half year or one year consumer shopping model data gather period.

Because every kind of industry, consumers will continue to demand price and quality change , with a wide range of convenient fulfilment options among of different kinds of products or services supply. Overall, the (AI) big data gather procedure gives opinion concerns every time retail experience will become more exciting, simple and convenient, depending on the consumer's ever-changing needs. So, I believe that (AI) big data gather every conclusion or result will be different, due to consumer's price and quality demand will often change to every kind of product or service supply in retail industry. So, how to shape (AI) big data gathering's analytical conclusion or result more clear. I shall recommend organizations need to build great understanding of and a stronger connection to increasingly empowered consumers before they plan and implement how to apply (AI) big data gather tool to predict consumer behavior as below:

Firstly, (AI) is empowered by technology, the consumer is redefining value. The traditional measures of cost, choice and convenience are still relevant, but not control and experience are also important. Globally, consumers have access to more than 2 billion different products choice by a wide range of traditional competitors and dynamic new entrants, all experimenting with new business models and methods of client engagement.

As choice increases, loyalty becomes more difficult familiarity and the consumer becomes more empowered. Businesses will have no choice and constantly innovate and disrupt themselves by meeting new technologies of high standards and expectations of consumers. So, (AI) data gather tool will need to follow different target group of consumers' needs to follow their different kinds of product design or style choice preferable to gather data in order to conclude the different target groups of consumer behavior to give opinion more clear and accurate to let businessmen to understand more clear how its customers‘ behavioral choice trend in the future half month, even to two years period.

Secondly, businessmen need to adopt changing technologies rapidly. Technology will be the key driver of this retail industry. Industry participants will only success if they have a clear prediction to focus on how to using technology to increase the value added to consumers. They must , however, do so will I realistic assessment of their costs and benefits. Hence, (AI) big data gather technological tools will need to design to help them to gather data efficiently by these ways, such as the internet of things ( IOT), artificial intelligence (AI) machine learning, augmented reality (AR)/virtual reality (VR), digital traceability. So, future (AI) big data gather tool are predicted to be most influential customer behavioral positive emotion changing tool for retail , due to their widespread applications , ability to drive efficiencies and impact on labor in order to impact consumer behavior changing effort from negative emotion to positive.

Thirdly, (AI) big data gather tool is an advanced data science of consumer behavior predictive tool. Businesses will have to bring the journey from simply collecting consumer data to using it to scale and systematize enhanced decision making across the entire value chain. When focused on their business goals, industry players should not lose sight of the impact that future capabilities and transformative business models may have on society.

However, (AI) big data gather tool will encounter these challenges when any business plans and implements to apply it to predict consumer behavior in retail industry. The challenges include that as below:

1. The high cost and difficulty of implementing new technologies . The (AI) big data gather tool needs capital and capabilities to be designed to implement to be applied to different retail industry users. so, expensive barriers to innovation, an organization and the skillsets of its people to support a new design of (AI) big data gather tool, highly digital technology may be required.

2. Slow pace of cultural change. Consumers need to adapt or accept (AI) new technology consumption model in the traditional retail industry. The rate of change is outpacing the ability of businesses to keep up. (AI) big data gather tool needs to be designed to adopt in new or evolved business model requires, in most cases, a new level of customer behavioral predictive machine operation will impact to influence any retail businesses’ consumer

behavioral changes at a minimum, an organization's structure, capabilities, culture and decision making. If the retail business expects to apply (AI) big data gather tool to predict how to change its consumer behaviors and how their consumption behaviors will tend to change in order to achieve to change their positive emotion from negative emotion before they choose to buy its product or consume its service in success.

### 3.3 Challenge to using (AI) neural networks to predict customer behavior from big data gather tool

The challenge to using (AI) neural networks to predict customer behavior is similar to predict travelling consumers' entertainment need challenge. (AI) big data gather tool will encounter the challenge: How can predict customer behavior be represented as sequential data describing the interactions of the customer with a company or an (AI) data gather system through the time, e.g. these interactions are items that the customer purchase or views ? So, every customer data gather , (AI) needs to spend time to analyze how and why to cause whose consumption behavioral choice. It is too difficult matter or judgement for (AI) learning. So, (AI) needs to spend time to learn how to analyze every customer's shopping behavior or action in order to gather all different consumers' past shopping action information in order to help business owners to predict future its potential customer shopping behavior how to change more clear and accurate prediction.

(AI) big data gather tool needs to learn to know that how to judge every customer interaction likes purchases over time can be represented with sequential data. Sequential data has the main property that the order of the information is important. Many (AI) machine learning models are not suited for sequential data, as they consider each input sample independent from previous ones. Therefore, at the end of the sequence, (AI) big data gather learn machines need to keep in their internal state of every customer purchase data, kind of product or service, price , whole year consumption times form all previous inputs, making them suitable for this type of data.

However, consumer behavior can be represented as sequential data describing the interactions through the time. Examples of these interactions are the items that the user purchases or views. Therefore, the history of interactions can be modeled as sequential data, which has the particular trial that an incorporate a temporal aspect. For example, if a user buys a new mobile phone, who might purchase accessories for this mobile phone in the near future or it the user buys a electronic book or paper book , he might be interested in books by the same author. Therefore, to make accurate predictions is important to model this temporal aspect correctly. To solve this predictive challenge of consumers to buy the product. One count the number of purchased products of a particular category in the last N days, or the number of days since the last purchase.

So, the (AI) big data gather designers can attempt to produce a feature vector which can be fed into a machine learning algorithm such as " logistic regression" will be the main feature and function to any (AI) big data gather machine to learn how to apply this " logistic regression" function or feature to predict any customer behavioral change for any product purchase or service consumption to the (AI) predictive consumer behavioral business users. Every different kinds of product purchases or services consumption will be needed to design " different model of logistic regression" in order to follow the kind of business to predict whose consumer purchase or service consumption behavior to predict more accurate.

## 3.4
## Challenges of artificial intelligence, algorithms technology and machine learning impact to consumption market

The challenges of artificial intelligence, algorithms technology and machine learning impact to consumption market are similar to travelling entertainment consumption market. Markets have played a key role in providing individuals and businesses with the opportunity to gain from trade. If (AI) big data gather tool can predict how to change potential customer behavior in success. The challenges to consumers will face that the overall market consumption model will be dominated by the businessmen only. So, it is not fair or reasonable to consumers, because (AI) big data gather tool has controlled or dominated all consumers‘ minds and it has predicted how and why every kind of product or service consumer shopping model or consumption behaviors how will change.

It will bring this questions: How can market designers learn the characteristics necessary to set optimal, or at least better, reserve prices after they had gather all data to conclude the analytical results of their consumers behaviors how will change? How can market designers better learn the environments of their markets?

In response to these challenges, artificial intelligence (AI ) and machine learning are important tools for market design. For example, retailers and marketplaces , such as eBay, Amazon and many others are mining their vast amounts of data to identity patterns that help them create better shopping experiences for their clients and increase the efficiency of their markets. By having better prediction tools, these and their companies can predict and better manage dynamic consumption market environments. The improved forecasting that (AI) and machine learning algorithms provide help marketplaces and retailers better anticipate consumer demand and producer supply as well as help target products and activities for segmented markets. Another important application of (AI) 's strength in improving forecasting to help markets operate more efficiently is in electricity market example. To operate efficiently, electricity marker makers can attempt to apply (AI) machine learning tool to follow every household family electricity consumers' past electricity consumption record to judge ( predict) how it will be every family's forecasting in the year.

An inaccurate forecast in the electricity supply and demand that can dramatically affect electricity market bad supply outcomes causing high variance in electricity charge prices or worse, blackouts. By better predicting every family's electricity demand and supply , electricity market makers can better allocate power generation to the most efficient power sources and maintain a more reasonable electricity stable charge market. Any example is design market, the application of (AI) algorithms to market design are already widespread and diverse.

(AI) algorithms technology , it is a safe that (AI) will play a growing role in the design and implementation of market over a wide range of applications. The challenges are that how (AI) can guarantee accurate to predict when and why and how consumer behavioral changes to any retail industries. In fact, retailers will need to discover the value that (AI) can bring to what benefits to influence their customer behaviors.

In the future, (AI) will bring their benefits to influence travelling entertainment customers to build positive emotions to any travelling entertaonment providers in these aspects as below:

1. Future (AI) big data gather tool will be an area of compute science that deals with giving machines , the ability to seem like they have human intelligence. In short, it is the power of a machine to copy intelligent human behavior. For examaple, machine learning algorithms are being integrated into analytics and customer relationship management platforms to uncover

information on how to better serve customers, chat bots have been incorporated into websites to provide immediate service to customers.

2. (AI) adoption continue to rise with chat bots taking the lead. Due to increasing ease of deployment , instant availability and improved quality, chat bots will become more and more common to manage customer service queries and to make intelligent purchase recommendations. Also, retailers can engage this kind of technology to answer continue questions and supplement customer support with chat-based shopping experience. So, (AI) and declines personalized, customized and localized experiences to customers.

(AI) will be applied across the entire retail product and service cycle, firm manufacturing to post-sale customer service interactions. Hence, retailers can use (AI) to its fullest potential will be also to influence purchases in the moment and anticipate future purchases, guiding shoppers towards the right products in a regular and highly personalized manner.

3. (AI) technology can rise the conscious customers. Customers are demanding an increased interest in the ethical practice of the brands they buy from. Todays, customers have a well-developed sense of what is solely intended to drive sales. This has lead to a rise in consumers ho make values based judgements about what to buy and where to shop. These consumers believe their purchase habits have an impact on the world. To win customers, retailers need have good conscious to predict consumers' desire. Future, (AI) data gather technology will be a good consumer behavior predictive tool to predict about for years will now become customer expectations and will have drastically changed the path to purchase. So, (AI) data gather tool is the predictive consumer expectations tool on every interaction, they have these brands.

4. Future (AI) can be impacted to influence consumer behaviors by its potential to free up time, enhance, quality, and enhance personalization. The industries include: Healthcare industry can apply (AI) to support diagnosis by detecting variations in patient data, early identification of potential pandemics, imaging diagnostics; automat industry can apply (AI) to autonomous fleets to ride sharing, semi-autonomous features, such as driver assist, engine monitoring and predictive, autonomous maintenance; financial service industry can apply (AI) to design the suitable personalized financial planning, fraud detection and anti-money laundering and automation of customer operation; transportation and logistics industry can apply (AI) to autonomous trucking and delivery, traffic control and

reduced congestion and enhanced security; technology, media and telecommunications industry can apply (AI) to search media, and recommendation, customized content creation and personalized marketing and advertising to attract retailers to promote; retail and consumer industry can apply (AI) to design personalized production, anticipating customer demand, , inventory and delivery management; energy industry can apply (AI) to read and record smart metering , more efficient grid operation and storage and predictive maintenance; manufacturing industry can apply (AI) to enhance monitoring and auto-correction of processes, supply chain and production optimization and on-demand production.

Hence, future (AI) technology will impact consumer technology when any retailers apply it to assist its manufacturing processes or product sale or service provision processes to satisfy consumers' needs, it means that it can help any retailers to influence positive emotion to consumers in their whole sale or consumption or purchase proccesses.

5. (AI) and machine learning technologies make it possible to capture, process, and inter data on a massive scale effectively , then any human being could ever do. For example, Criteo's creative technology " Kinetic design" can apply insights from 1.2 billion monthly impressions to select and optimize individual branded advertisements components according to each shopper's preference and intent. This ensures more personalization and visually inspiring on brand ads. resulting in up to 12% more sales for (AI) technology advertiser clients.

Moreover, advertisers can now engage and inspire shoppers on a more personal level, rendering custom ads. it real-time for every impression. So, designer continues to learn from each design's success to make ads. more and more effective over time. Furthermore, brands are increasingly using paid search on retail sites to draw attention to their products on the crowded online shelf, e.g. Google shopping is a key growth area's more users are engaging with shopping ads. and across the globe. Google shopping has become essential to retailers' marketing strategies, but is a difficult channel to apply its tool to be promoted effectively . Thus, future (AI) and machine -learning technologies can dramatically improve digital commerce performance application to apply (AI) and machine learning to digital consumer. So, future (AI) technology can be applied to digital commerce aspect, it will fall into the categories of pattern recognition, classification, prediction and consumer behavior.

In conclusion, the benefits of using (AI) in digital commerce include: improved efficiency in discovering the relationships between datasets over traditional methods, which require complex modeling and coding, improved accuracy for clearly defined processes that involve a lot of manual processing, ability to deal with a large emotion of data with many attributes, for example: customer behavior data, multichannel and multi-device data , complex product data and fraud detection, more accurate analysis, such as customer segmentation sentiment, analysis and personalization frequent algorithum refreshes, such as several times a day, to capture the changes in customer and market behavior.

Finally, however, a lot of types predictive consumption behavior around (AI), in particulars that driven by vendors claiming their solutions are (AI) , ready and can deliver dramatic improvements over existing technologies. Application leaders for digital commerce can be misled into believing that (AI) can solve all their problems, which is not true for n in-depth discussion of the (AI) consumers and market behavioral predictive tool and machine -learning technologies bot. Thus, (AI) prediction consumer behavioral technology can give beneficial quantitative analysis for forecasting in business and market especially in consumer behavior and in the consumer decision-making process ( consumer choice model) more effectively and efficiently.

## Is Artificial Intelligent the most effective and accurate travelling consumer behavioral tool?

Is (AI) the best and the most effective and accurate travelling consumer behavioral prediction tool to compare other kinds of consumer behavioral prediction tools? Nowadays, travelling entertainment providers' competitions are serious businessmen often find different kinds of methods to attempt to predict travelling consumer travelling entertainment need of changes. The travelling consumer behavioral predictive methods can include as these below methods, instead of (AI) big data gathering tool.

Firstly, statistics is the popular mathematic method, it applies auto-regression, liner regression, structural equation modelling, logistic regression statistic techniques to be used to predict consumer behaviors. Secondly, it is classification method, it sis a support vector machine to assist businessmen to make consumer behavioral prediction, it also includes decision making tress diagram technique. Thirdly, it is rule mining method, it is algorithm, market base analytic etc. business marketing concept analytical tool, it also includes graph mining technique tool. Next, it is psychological prediction model tool, it is psychology prediction model too, it is a kind of psychological method to predict consumer behaviors. Finally, it is the most updated and potential artificial neural network (ANN) machine tool, it gathered big data, then it will carry on analyzing and applies psychological method to conclude the most accurate and reasonable solutions to give recommendation to businesses to predict when and how and why their consumer behaviors will change. So, it is one owned human mind's machine and owned psychological and analytical efforts to replace humans to make any judgement in order to make the most accurate predictive behavioral changes for consumers, instead of the traditional marketing concept and psychological and mathematic methods to predict consumer behavior, (AI) big data gathering tool will be another new tool.

What are the advantages of (AI) tool to be used to predict consumer behaviors as well as what are the different between it and other traditional consumer behavioral predictive tools? I shall explain as below:

Firstly, as above all case studies are explained to (AI) questionnaire design method benefit, I believe (AI) big data gathering tool can be applied to help human to analyze and design any the suitable valid questions to enquire any kinds of business consumers in order to gather the most meaning and useful opinions to conclude the most accurate consumer behavioral prediction for every questionnaire. So, future (AI)'s analytical effort and decision making effort most be exceed above human's judgement efforts. So, future (AI) can help human to design the most useful and meaning different kinds of valid questionnaire ( survey) questions as well as assist humans to analyze and make accurate decision making and conclusions to give opinions to help businessmen to predict when consumer behaviors will change and how their consumption behaviors will change to influence their businesses in order to help them to make any efficient and effective and accurate solutions to avoid consumer number to be decreased and the most important benefit is that it can give opinions to help businessmen to explain why ( what the factors ) cause their consumer behaviors change suddenly. It will be human's efforts can not achieve to exceed (AI)'s efforts in the future.

Secondly, (AI) can make artificial machine judgement and analytical effort, without human misleading or unfair or unreasonable judgement. So, it can make more fair and reasonable and accurate conclusion to give opinions to predict when, how and why consumer behaviors will change suddenly to the kind of business in customer model building process and evaluating the results of customer relationship management –related investment more accurate.

Furthermore, (AI) big data gathering tool will help businesses to improve the success rate of acquiring customers, increasing sales and establishing competitiveness. (AI) big data gathering tool can give opinions how to build customer loyalty to be positive emotion impact and it can find solutions to avoid every client's negative emotion causes to bring complaints behavior to the businessman's product or service. For example, Telecom industry and aggressive research has been conducted in this by applying various data mining techniques to avoid long distance phone call users' complaints. If gathered any long distance phone call users' past complaint data to record what are their general complaint issues. Then, (AI) tool will analyze all these past complaint issues to conclude and give opinions to let Telecom knows whether which aspects encounter challenge that Telecom needs to improve it's long distance phone call services or functions in order to satisfy Telecom's long distance phone call users' needs for long term. After Telecom attempted to improve its services and/ or functions from (AI) opinions and solution methods, when it fell it's long distance phone call users have positive emotions to satisfy its service performance and function performance. Then, it can prove (AI) tool's opinions and solutions are useful. The consequence is that their complain numbers will be decreased and they won't plan to choose another long distance phone call telephone service company to replace Telecom long distance phone call service more easily.

So, (AI) big data gathering tool can concentrate on finding focus on components of customer relationship management method and datasets more accurate and efficient and effective than human's data gathering and analytical effort. It implies (AI) big data gathering tool has unique more efficient and effective and accurate dataset gathering and analytical and judgement and decision making effort, it is human can not achieve.

Thirdly, (AI) big data gathering tool has much customer loyalty predictive effort. It's effort is more easily subsequently selected, reviewed and classified to compare human's gathering data effort in whole data gathering and analytical process.

In (AI) big data gathering process, (AI) can organize whole big data gathering process and technique more easily in short time. It will include these four steps. The first stage is that customer identification stage, customer identification also known as acquisition has to do with targeting the population , who are most likely to become customer segmentation. So, (AI) can help different kinds of businesses to gather their competitors' consumer purchase behavior data in short time, it is human can not achieve. The second stage is that customer attraction stage, after (AI) maker has been segmented for the business when it has ensured to gather the businessman's global competitors' consumers data. Then it analyze these all data to find solutions / methods to give the best opinions to the organizations how to achieve the direct effort and resources into attracting the target customer segments. The third stage is that customer retention, it can be defined as the activity that an organization undertakes in order to reduce customer defections. TO be successful, customer retention starts with the first contact on organization has with a customer and continues throughout the entire lifetime of a relationship involves loyalty programs, one to one marketing and complaints management. SO, (AI) can consist the business to find the best or the most reasonable , efficient , effective solutions or methods and it will conclude all these solutions to find the most reasonable and useful opinions to achieve to the aim to help the business to reduce customer complain numbers and help the business to build confident loyalty relationship between it and its clients. SO, (AI)'s analytical effort and decision making effort can be more accurate than human's analytical effort and decision making effort. IT can achieve it's consumer behavioral predictive aim more accurate and efficient and effective in the shortest time to compare human.

Fourthly, (AI) big data gathering tool can design more accurate dataset program for questionnaire (survey) to compare human's questionnaire ( survey ) effort. It means that (AI) can spend less time to research and make judgement what are the most reasonable and meaning questions for different kinds of businesses' needs. This includes data conduction a questionnaire, survey or interview of the individual or environment researched, public data repository: This includes commercially available public data; organizational data; this contains data collected from an organizational database, organizational information system. For example, their website log details etc. It also includes company transactional data, data purchased from a company.

For example, one vehicle sale company expects to research all global vehicle sale companies' past the different kinds of vehicle styles, design sale number data, the different kinds of vehicle style, design sale price data, every country's vehicle consumer number to the vehicle purchase number data to the vehicle company in short time. (AI) big data gathering tool can help the vehicle sale company to gather all any one for these global vehicle sale competitors' past data in the short time. It is human effort, who can not achieve this efficient, effective and accurate data gathering aim for this vehicle sale company. Even, when (AI) had gathered all global it's vehicle competitors' past sale data, (AI) can make more accurate analytical and judgement and decision making effort to design different kinds of questionnaire ( survey) questions to prepare to enquire it's different target segmentation vehicle potential clients in order to predict what are their needs to choose to buy any vehicles from the vehicle company. SO, (AI) tool can conclude more accurate conclusions and give the most reasonable and useful opinions to let the vehicle company to know in order to predict what are it's potential vehicle buyer's needs and manufacture the suitable vehicle styles or designs to raise their vehicle purchase desires.

Fifthly, (AI) tool is only one perfect tool for big data gathering in order to achieve accurate results and increased profit. What is (AI) big data gathering mean? The term " big data"gathering describes the accumulation and analytical of vast amounts of information, but big data is much more than a big amount of data. It is also the ability to extract meaning to sort through big volumes of numbers and find the hidden patterns, unexpected correlations and surprising connections that can be used in different industries like medical field, security and protection field or marketing that adopt " big data driven" decision making enjoy significantly greater productivity than those that do not. So, the benefits of (AI) is given to the company by using big data repaid complexity of implementation projects and hence project risks, when accelerating time to value. It is why that human's gathering effort can not replace (A I) data gathering effort.

All analysing above benefits to (AI) big data benefits to any organizations, it brinfs this question: How can (AI)apply big data gathering and analyzing to predict when and how any why consumer behavior will change suddenly? The purchase decision making process is consumers reducing purchase choice behaviors.

Travelling consumers are being considered pure rational beings ( consumer tried only to satisfy self-interest). Hence, due to future (AI) owns human's psychological , analytical , emotional predictive, purchasing decision making effort.

(A I) will be assumed to sees one travelling customer how who will make travelling destination and travelling package decisions. So, after the (AI) gathered all data concerns the find of business's past travelling customer segmentation travelling destination and travelling package design activities, e.g. age, sex,. income level, the different kinds of travelling package sale number, the different kinds of travelling package price variable sale etc. different kinds complex data.

It can make more accurate psychological and analytical effort to predict when the travelling provider's travelling consumer behaviors will change as behaviors will change as well as find what reasons their travelling package choice behaviors will change and how trend of their travelling consumer behaviors will change more accurate. For example, today there are a lot of industries that use big data: healthcare (treatment ) becoming personalized and patient centric and predictive analysis are used to prevent diseases for example Angelina Jolie underevent a predictive double mastectomy after learning she had 87% rich to developing breast cancer, sports ( by using sensors data are collected from players during a game in order to improve their playing schemes), weather(more than 60 years of global weather analysis are used to predict the risk of future extreme events), logistics ( smart tucks and smart species, agriculture (monitoring weather and soil conditions for optimum point of harvesting).

Consequently, due to the evolving consumer demands, and the ever growing digitization, the world is digitally transforming which means the new technologies are needed to be used and driven significant business improvement. So, such as why (AI) tool will be our future main predictive tool to help travelling service providers to predict when, how and why their potential travelling customer package choice behavioral will change. Big data is one of the our channels through digital transformation is made, together with cloud, mobile and networks. The challenges for digital transforming and therefore using A I
big data gathering tool as main technology are: digital proficiency, legacy systems, security and jobs becoming absolute.

In the future, big data can use data from text to travelling photo picture , sounds, movies, musics satellite coordinates or any other type of input or output data that type of input or uouput data that came from different influential aspect. It is cloud solutions, bring big data will be for predict insight driven by travelling business stategy, differnet kinds of travelling package design strategies and new travelling consumer relationship, predictive travelling consumer's travelling package design behavioral strategy. Using the right data in the right business decision will mean smart decisions, new opportunitites and utimately a big competitive advantage.Hence (AI) big data gathering tool is different is that (AI) can be one depth in-memeory database function, it can make real-time data analytics that provide meaningful information in short time, it is also the visualization tool , such as SAP Lumira, allow this exploration and understanding of the data, and ultimately supports the decision making process. All above these features, which will be human's data gathering effort who won't exceed (AI) big data gathering effort. Hence, future (AI)big data gathering will be the best choice to assist travelling service businesses to predict travelling consumers' travelling package design behaviors successfully.

Reference

Adrian, P. (2012). Introduction to marketing theory & practice, 3 rd edition, London: Oxford press.

Ajzen, I (1991). The theory of planned behavior. Organizational behavior and human decision processes, 50(2), 179-211. doi: 10.1016/0749.5978 (91) 90020-7.

Alba, Joseph W. and J. Wesley Hutchinson (1987). " Dimensions Of Consumer Expertise", Journal of consumer research, 13 March, 411-454.

Bailey, L., Mokhtarian, P.L. Little, A. (2008). The broader Connection Between Public Transportation, Energy Conservation And Greenhouse Gas Reduction, Report Prepared As Part Of TCRP Project J-11/Tasks Transit Cooperative Research Program, Transportation Research Board Submitted To American Public Transportation Association in http://www.apta.com/research/into/online/land_use.cfmi, accessed 17 April 2008.

Baucer, R,"Consumer Bhavior As Risk Taking , In Risk Taking And Information handling In Consumer Behavior", D. Coxceds Harvard University Press, Cambridge, Mass 1976.

Biederman, P. (2008). Travel and tourism, Pearson Prentice Hall, New Jersey.

Bogers, R. P., Brug, J. Van Assema, P., & Dagnetie, P.C. (2004) , Explaining fruit and vegetable consumption: The theory of planned behavior and misconception of personal intake level. Appetite, 42,157-166.

Bolton, Ruth N. (1998), " A Dynamic Model Of The Duration Of The Customer's Relationship With A Continuous Service Provider: The Role Of Satisfaction", Marketing Science, 17 (1), 45-65.

B.Shiv and A. Fedorikhin, " Heart And Min In Conflict: The Interplay Of affect And Cognition In Consumer Decision Making", J. Consumer Res., vol. 26, pp. 278-292, Dec. 1999.

Brown, K.W., Ryan, R.M. Reswell , J.D. (2007). Mindfulness: Theoretical Foundatins And Evidence For Its Salutary Effects. Psychological Inquiry, 18, 211-237.

Burke, R.R. : Behavioral effects of digital signage, J. Advertising Res. 49(2), 180-185 (2009).

Cant, M., Brink , A. & Brijall, S., Consumer behavior, Cape Town, South Africa: Juta, 2006.

Conner, M. & Abraham, C. (2001). Conscientiousness and the theory of planned behavior: Toward a more complete model of the antecedents of intention and behavior. Social psychology bulletin, 27, 1547-1561.

Cooper C. Mallon, K, Leadbetter S, Pollack L, Peipins ( 2005) , cancer internet search activity on a major search engine, United States 2001 to 2003, J Med Internet Res. 7(3): e36.

Cope, R. R. Cope and H. Davis (2008). Disney's virtual Queues: A strategic opportunity to co-brand services ? Journal of Business & economics research, vol. 6 no10, 13-20.

Cornelia, B.F. (1999) Rural development news, the North Central Regional Center For Rural Development vol. no 24 , IOWA.

Couper, M.P. J. Blair and T. Triplet ( 1999). A Comparison Of Mail And E-mail For a Survey Of Employees In USA Statistical Agencies. Journal Of Official Statistics, 15, 39-56.

David J. Nowak & Gordon M. Melsler (2016) " Air quality effects of urban trees and parks." National recreation and park association, USA.

Data monitor ( 2008). The proctor and gamble company. Retrieved Nov. 15 2009 from http://www.datamonitor.com/

De Hollander, A. E. M., J.M. Melse, Elebret & P. G.N. Kramers ( 1999), " An Aggregate public health indicator to represent the impact of multiple environmental exposures" Epidemiology: 606-617.

De Visser, R.O., & McDonnell, E.J. ( 2013). " Man points": Masculine capital and young men's health. Health psychology, 32( 1), 5-14. doi:10. 1037/ a0029045.

Dunn, J & A Neumsister (2002). Knowledge management in the Information age. E. business review, Fall , 37-45. Jounral of service, spring 2011, vol. 4, no1, De Grovte (2009).

Dyer, D., F. Dalzell & R. Olegario ( 2004). Rising tide. Lessons learned from 165 years of brand building at Procter and Gamble. Boston, MA: Havard Business School Press.

Eysenbach G (2006) Infodemiology: Tracking flu- related searches on the web for syndromic surveillance. American Medical Informatics Associaion Annual Symposium Proceedings , Curran Associates, Red Hook, NY, pp. 244-248.

Ettredge M, Gerdes, J. Karuga , G (2005) Using web- based search data to predict macro-economic statistics. Commun ACM 48: 87-92.

Felce, D. and Perry, J. (1995). Quality of life: A contribution to its definition and measurement, vol. 16, no.1 pp: 51-74.

Feldman, Jack M. And John G. Lynch Jr. (1988), "Self-
Generated Validity And Other Effects Of Measurement On Belife, Attitude, Intention And Behavior", Journal of applied psychology, 73(3),421-35.

Fiese, M, Hofmann, W., & Wanke, M (2009). The impulsive consumer. Predicting consumer behavior with implicit reaction time measurement. In M. Wanke (ed.) Social psychology of consumer behavior (pp.335-364). New York, NY: Psychology press.

Fitzsimons, Gavan, J. And Vicki G. Morwitz ( 1996), " The Effect Of Measuring Intent On Brand-Level
Purchase Behavior", Journal of consumer research, 23 (1), 1-11.

Hallerman , D. (2008) video Advertising Online: Spending And Pricing , New York. E-Marketer.

Harriet Griffey. (2010) The art of concentration, enhance focus, Reduce, stress and achieve move. Macmillan publishers ltd,Basinastoke and Oxford, London UK.

Helleman, D. (2008) Video Advertising Online: Spending And Pricing , New York, E-Marketer.

Hensen, C. (2003). Kreuzfahrtourismus.www.christoph- hensen.de/ Facharbeit.pdf.

Huang, H.I. (2012). An empirical analysis of the strategic Management of competitive advantage: a case study of higher technical and vocational education in Taiwan ( Doctoral dissertation,
Victoria University).

Jamieson, Linda F. And Frank M. Bass ( 1989), " Adjusting Stated Intention Measures To Predict Trial Purchase Of New Products: A Comparison Of Models And Methods," Journal of marketing research, 26 ( August), 336-45.

Korea Ministry Of Environment. Public Organizations spend 2.2 Trillon Korean Won To Purchase green Products in 2014; Ministry Of Environment: Sejoung, Korea, 2015.

Kremers, S.P. J., De Bruijn, G.J., droomers, M., Van Lenthe, F. J., & Brug, J. (2005). Environmental interventions for selected dietary behaviors in adults. In J. Brug & F. J. Van Lenthe ( eds.) , Environmental determinants and interventions for physical activity, nutrition and smoking: A review pp. 282-315. Rotterdam: Erasmus Medical Center.

Lee, D.; Kim, M. ; Lee, J. adoption of green electricity policies: Investigating the role of environmental attitudes via big data-driven search-queries. Energy policy 2016. 90, 187-201.

Lee, Terrence, " Tech in Asia-connecting Asia's startup system " Tech. in Asia-connecting Asia's startup ecosystem, N.p.,4 July 2016.

Los Angeles Country Department Of public Health (2016), Country Health Ranking Model, Retrieved From
www.countryhealthrankgings.org/our-approach. USA.

Mayne, Lonnie. " Evolve of die in the age of the consumer". Entrepreneur, N.P. , 16 Apr. 2014. web of Oct. 2016.

McGregor, S.L. T., & Goldsmith, E.B. (1998). Expanding our understanding of quality of life, standard of living and well-being. Journal of family and consumer science, 90(2), 2-6, 22.

McMichael, A.J. M. Mckee, J. Shkolnikov and T. Valkanen ( 2004), " Morality trends and setbacks, global convergence or divergence?", Lancet 363, 1155-1159.

Melse, J.M. & A.E. M. De Hollander (2001). " Human Health And The Environment", background document for the OECD Environmental Outlook, OECD, Paris.

Moschis, George p. & Roy, L. Moore ( 1979), " Decision making among the young. A socialization perspective " Journal of consumer research , 6 ( September).

Mulligan, M. Banerjee, T & Thomas, N. (2008) ,European Paid Content And Activity Forecast, (2008 to 2013), Jupiter Research.

Peter, J., Ryan, M, M, " An Investigation Of Perceived Risk At The Brand Level, " Journal of marketing research, 13 May 1976, pp. 184-188.

Pieters, R., & Wedel, M. (2007). Goal Control Of Visual Attention To Advertising: The Yarbus Implication. Journal Of Consumer Research, 34, 224-233 ( August).

Parasuaman, and Leonard L. Berry (1985), " Problems And Strategies In Sevices Marketing", Journal of marketing, 49 ( Spring), 33-46.

Priesnitz, W. (2007) Counting Our Food Miles. Natural Life, 1 July.

R.C. Oliver, " When is consumer loyalty?" J.Marketing vol. 63, pp.33-44.1999.

Reggiani, A . (ed). 1998, accessibility, trade and locational behavior, Ashgate publishing ltd, England.

Rushe, D. (2013) " The 10 best paid CEO in America". The Guardian , 22 Oct, ( online). Available at:
http://www.theguardian.com/business/2013/Oct22/best-paid-chief-executives-america (Accessed: 3 May 2014).

Spiekermann and Wegener (2007), update of selected potential accessibility indicators. Final report, urban and regional research ( S&W), RRG spatial planning and geoinformation. ESPON. Available online at http:// <www.espon.eu/mmp/online/website/ contentprojects/947/1297/file_2724/espon_accessibility_update-2006-fr_070207.pdf>, accessed on 1 July 2009.

Starbucks (2014) Our company available at http:// www. starbucks.com/ about- us/company-information ( accessed: 3 May 2014).

Shostack, G. Lynn ( 1984), " Designing Services That Deliver", Harvard Business Review, 62 ( January-February), 133-9.

Shostack, G. Lynn (1985), " Planning The Service Encounter ,in the service encounter" , John A. Czepiel, Michael R. Solomon, and Carol F. Suprenant, eds. New York: Lexington Books, 243-54.

Shostack, G. Lynn (1987), " Service Positioning Through, Structural Change", Journal of marketing, 51 ( Janurary), 34-43.

Soloman, Michael R. (1985), "Packaging The Service Provider", Service Industries Journal , 5(1), 64-71.

Stevens, C.W. (1980), "K-MartStores Try New Look To Invite More Spending" The Wall Street Journal, Nov. 26, 29-35.

Sullivan, Nicholas P(2007). You can hear me now: How Micro loans and cell phones are connecting the world, San Francisco, CA: John Wilsey & Sans, 2007.

T. Ambler, A. Ioannides, And S. Rose, " Brand s On The Brain : Neuroimages Of Advertising ", Business Strategy rev., vol. 11, 3. pp. 17-30. 2000.

Westbrook, Robert A. ( 1980), " Intrapersonal affective influences on consumer satisfaction with products, " Journal of consumer research , 7 ( June) 49-54.

Wiig, k.(1993). Knowledge management foundations: Thinking About thinking. How people and organizations create, represent and use knowledge vol.1 , of knowledge management series schema press: Arlington, TX.

World Health Organization (2003). Diet, nutrition and the prevention of Chronic diseases report of a joint WHO/FAO. expert consultation. Geneva: World Health Organization.

Wysocki, B. (1979), " Sight, Smell, Sound: They're all arms in retailer's arsenal" The Wall Street Journal, Nov. 17, 1979. 1-35.

Yale Center For Environmental Law And Policy (2006). Environmental Performance Index. Data available on-line at http://epi.yale.edu

How can artificial intelligent tools predict travelling consumer behavior in airline and air agent travelling market ?

I believe that applying (AI) big data tool to predict vehicle buyer consumption choice behavior, it is similar to predict traveler consumption choice behavior. In this chapter, I shall indicate how to apply (AI) big data gathering tool to predict vehicle buyer consumption choice behavior. Then, I shall its what its similar points to be applied to predict traveler consumption choice behavior.

Nowadays, many vehicle manufacturers hope their vehicles can attract to vehicle buyers to choose to buy their vehicles. However, there are many different brands of vehicles to provide to them to choose, so the vehicle market competition is very serious.

How to judge their different kinds of vehicle price which is reasonable acceptance to attract vehicle buyers to choose to buy the brand of vehicle manufacturers' any kinds of vehicles, e.g. fast speed sport style vehicles, comfortable and slow speed common cars, for four passengers common small size or more than four passengers common large car size?

How to evaluate the vehicle prices issue is important factor to influence vehicle buyers' choices. Either if the brand of vehicle price is too high to compare other brands of similar vehicle price, it will influence many vehicle buyers choose to buy other brands' vehicles or if the brand of vehicle price is too low, it will influence vehicle buyers feel this brand's vehicle machine quality or safe driving level or manufacturing steel material or speed or not comfortable sitting etc. different factors is worse to compare to other vehicle brands' similar vehicle products.

Thus, if the brand of vehicle manufacturers can predict how to design vehicles which can attract many vehicle buyers to choose to buy whose any vehicle products. What are future vehicle buyers' favorable vehicle styles? Then, the vehicle manufacturer can concentrate on manufacturing the kind style of vehicle products to sell already. It will reduce its vehicle manufacturing investment risk.

How to apply (AI) tools to predict vehicle buyers' behavioral consumption model? Whether artificial intelligent tools can predict automotive buyers' behavioral consumption model and predict future vehicle design trend. In fact, automotive brands and dealerships are facing an increasingly competition when attempting to manually gathering the vast quantities of data required to create customer focused programs that increase retention, ultimately new sales and service automotive business.

Building a based on that client's intrinsic needs and interests to any kinds of automotive vehicles at any given time. This is especially true in the automotive industry where the time span between purchases is measured in years. Because vehicle buyers would not like often to change their old vehicle to another new one. So, their decisions to buying another new vehicle, the time is usually after one year, even longer time. Hence, it seems any vehicles won't be frequent consumption products to the owned at least one vehicle family consumers (vehicle buyers). It implies that why vehicle manufacturers ought need to spend time to predict future vehicle buyer design choice for whole year vehicle buyer number growth because they won't often change preferable vehicle design to change another new vehicle more easily.

Hence, how to predict vehicle consumers' taste or preferable which styles of vehicle choices issues is very important. If the vehicle manufacturers can not manufacture any attractive vehicles to sell easily in this year. Then, it will lose time, money in this year because it won't know when the owned least one vehicle users or non-owned any vehicle users who will decide to buy one new vehicle or change another new vehicle ensure. The different brand vehicle dealers will possible wait more than one year to attract them to buy their vehicles if their styles are not attractive to compare other brands of vehicle competitors.

However, artificial intelligence and machine learning can help any vehicle manufacturers to find solution to solve patterns in highly to solve patterns in highly complex data-sets that are beyond the capability of a human brain, and then building and automatically acting on the customer insights it generates.

Given the automotive customer need for individualized communications, this technology is positioned to become a critical component of any successful vehicle retailer's domestic or/and overseas vehicle markets. How can vehicle manufacturers and retailers use (AI) to enhance their vehicle marketing campaigns? How will (AI) affect their vehicle sale marketing strategy? What criteria would they use when selecting on (AI) solution?

Vehicle consumers today are able to quickly access different brands of vehicle information, research vehicle products and reviews, negotiate prices and compare one vehicle brand or retailer to another resulting of the brands of vehicle customers. At the same time, the rise of " big -data mining", wearable devices that track user's every move and preference and greater contextualization in advertising and social media has resulted in consumer expectations of individualized. Thus, it seems that (AI) tools can be used to gather " big-data" and then they can make human's mind to analyze how to design kinds of vehicles to satisfy vehicle buyers' needs.

As automotive vehicle marketers can apply (AI) tools to achieve messaging strategies to meet the needs of this new generation of informed vehicle consumers, using data from a variety of sources to move from a variety of sources to move from mass- messaging to more personalized messages aimed at particular vehicle buyer segments, e.g. fast speed sport vehicle buyer segment, slow speed comfortable small size or large size of buyer segment. However, when 90% of vehicle marketers believe having a single vehicle buyer view is important, only 6% have achieved it.

However, one of the main issues vehicle marketers are facing the lack of capacity to efficiently sift through and analyze the massive vehicle buyer amounts of data required to create vehicle buyer individualized vehicle customer experiences easily. This is especially difficult for automotive dealers, the long periods between purchase cycles, and the highly considered nature of the vehicle purchase means that each vehicle dealer needs to not only track a large number of potential vehicle customers for an extremely long period of time, but each of those vehicle customers will generate a huge amount of different kinds of vehicle behavioral consumption data as they research their next vehicle purchase. However, by choosing the right (AI) technological tools and programs , vehicle dealers can solve this big data gathering challenge into a major advantage.

For Forrester vehicle brand example, vehicle consumers have more power over the Forrester vehicle brand's reputation than ever before. Mayne, L. (2014) indicated that Forrester calls this new (AI) tools is the " age of the vehicle customer", a 20 year business cycle in which the most successful vehicle enterprises will reinvent themselves to systematically understand and serve increasingly powerful vehicle consumers. To win in this new age, Forrester declares companies must become vehicle customer obsessed and the only sustainable competitive advantage is knowledge and engagement with customers, such as (AI) gathering data knowledge.

Thus, the biggest challenge vehicle businesses currently face is not the collection of a large quantity of vehicle consumer data, but what to do with that data once they have it. Even at a large vehicle data research firm, the data sets are often too big for a single analyze, or even a team of analysts to sort through and draw conclusion from. However, enter artificial intelligence and machine learning , an efficient technology solution that can continuously find patterns in highly complex data sets that are way beyond the capacity of a human brain and then automatic drive action based on the customer insights is generated.

What is (AI) machine learning tool? Machine learning is a type of (AI) that learns from data and is not explicitly program. Think Amazon, face book. Machine learning serves up relevant content based on an individual vehicle purchase behavior and experiences. More simply, machine learning is a computer program that can learn relationships between data, subject those learnings to errors functions, and then learn from its errors. The program in effect, trains itself.

Lee, T. (2016) explained that "Thus, (AI) tools can learn deep a more advanced branch of machine learning inspired by how our brain's nervous function, has also been found to be especial effective in identifying patterns from

data."
When this way sound is complicated from a vehicle dealer perspective, the implementation of a marketing program driven by artificial intelligence can take care of these tasks in an automatic vehicle fashion with little to no manual intervention required from the staff at time vehicle stores.
In practice at a vehicle dealership, the program will continue track vehicle customer behavior online, merging that data with any offline source ( like CRM or DMS data) and then analyze this aggregated vehicle buyer data set to predict what vehicle customer may be shopping for and what information they might like to relevance from different kinds style of vehicle design photos.

1.1
Why does travelling market seem to similar to vehicle market which can apply (AI) learning tool to predict travelling
consumer behaviors?

Artificial intelligence refers to complex in vehicle market and travelling entertainment market which is very seem to be applied to predict consumer behaviors.
(AI) machine learning that posses the same characteristics of human intelligence and that have all our sense, all our reason and think just like human vehicle buyer who prefer vehicle purchase choice or travelling consumer who prefer travelling package or travelling destination and airline choice. Besides, machine learning is the practice of using algorithms to collect and examine data, learn from it, and then make a determination or prediction about something in the world.
So, it can be attempted to gather data concerns that travelling consumer past travelling destination choice and air ticket price choice and different travelling package, e.g. high, middle, or low class hotel and foods supply and entertainment places choice in their past travelling journeys.
The machine is " trained" using large amounts of data and algorithms that give it the ability to learn how to automatically perform a task with increasing accuracy. Otherwise, deep learning is primarily based on artificial neural networks inspired by our understanding of the biology of human's brains.
Thus, (AI) big data can gather all these past traveler consumption behavioral choice data to make reference to analyze whether how many travelers will choose to go to the specific travelling destination in any time by the past

traveler number record to different travelling destinations, then it can gather the past air ticket sale price to different destinations and past travelling package design to different destinations in order to analyze whether it is the cheap airline ticket price factor or attractive travelling package factor or attractive travelling entertainment etc. in order to predict which factor is the most potential influential factor to they choose to go to the destination to travel in different time within one year. Then, traveler agent or airline can collect these big data to judge how to design their package to attract travelers to go to anywhere to travel or what the main factor influence most of them to choose to visit the destination to travel.

For example, travel agents or airlines can apply "Deep learning" breaks down tasks in ways that enables machines to assist them to predict when travelling consumer choice will be changed and why their travelling choice will change and how their travelling choice will change with increasingly complex tasks.

So, such as why (AI) technology can be applied to predict how travelling consumer behavior changes to bring to judge whether anywhere will be many travelling consumers who will prefer to choose travelling hot destinations next year or next month.

Then, travel agents and airlines can gather overall past travelling consumer data to analyze and conclude the more accurate prediction of different travelling destinations to the number of traveler. Then, they can choose how much air ticket price is more reasonable to charge to the travelling destination or how to design the travelling package which can bring more attractive to the prediction number of different travelling destination travelers in order to achieve to raise the different travelling destination number next year.

Thus, (AI) big data machine learning can help airlines or travel agents to solve how to design any attractive travelling package challenge. A travelling package is both one of the most important and carefully considered travelling entertainment consumption the majority of travelling people will ever make in their lifetime at least one travelling time.

It is also a prediction how travelling package will be designed that tends to be fundamentally tied to a travelling person's travelling destination choice identify and travelling package view of themselves. As the same time, travelling consumers' travelling choice changing lifestyles result in changing travelling destination needs, e.g. the country's young travelers can choose to change non-extreme exciting travelling entertainment package from past extreme exciting travelling entertainment package. Due to personal feeling factor in general. However, I believe that (AI) big data can also be attempted to predict when the country's young travelers will choose to change non-extreme exciting travelling behavior.

It is similar to automotive dealers need to remember that vehicle customers and prospects are individual human beings with risk, complex and ever-changing lives factors, these factors will influence every vehicle consumer why who feels has vehicle purchase need, and how who choose to buy the first vehicle if who decided to buy the first vehicle.

It seems that travelling agents or airlines need to remember that travelling consumers and different features or designs are very traveler beings with risk, complex and ever-travelling package attitude personal changing factors in different travel season, these factor will influence every individual traveler why who feels has travel entertainment need, and how who choose to buy different feature or design travelling package if who decide to travel.

The (AI) big data technological travelling customer behavioral prediction tool seems to be the best travelling behavioral prediction tool in the world are those that know every one of different country's traveler need. Their likes and dislikes which style of travelling package, preferences and travel destination changing tastes to travelling destination choices.

The capacity of the human brain, however, limits us from achieving these different type of travel package sales. In this competitive travelling destination choice entertainment environment, (AI) big data machine learning enables platforms to assist the air ticket and travel package sales team by tracking the travelling consumer behaviors of each travelling customer, learning and memorizing their preferences and predicting their future travelling destination choice and travelling package design needs.

Finally, I recommend that for a travel agent or airline travelling marketing platform to make their travelling customer engagement efficient and fully-functional, I should be able to: applying (AI) tools to track every travelling customer behavior across the web, connecting to a society of data sources, CRM, DMS, third-party, web travelling brands, social traveler email, click etc., aggregating and accurately cross-reference data from a variety of sources, leveraging this data to drive insights on a mass scale, as well as on an individualized basis, driving actions and automatically direct travelling customer engagement via multiple channels based on where each customer is in their travelling individual lifecycle.

## 1.2
## Why is (AI) big data gathering tool better than psychological and survey methods to predict traveler individual travel choice behavior?

Prediction travel behavioral consumption from psychology and survey methods.

How to predict travel consumption? It is one question to any travel agents concern to use what methods which can predict how many numbers of travelers where who will choose to go to travel more accurately. I think that who can consider how to predict travel behavioral consumption from psychology and survey travel choice prediction method, but it is better to apply (AI) big data gathering method to predict travel consumer's destination choice more accurate. The reason is as below:

The first reason is that traveller individual travel psychological desire is difficult to predict accurate more than (AI) big data gathering method, it is due that the data is past traveler's destination choice and travel package and ticket price actual data from (AI) big data gathering method. Otherwise, survey investigation is only traveler psychological thinking method. It lacks enough past actual traveler data gathering.

The second reason is that on the weakness of traveler individual psychological thinking view of survey investigation. It has evidence to support the relationship between self-identify threat and resistance to change travel behavior to any travelers, controlling for whose past travelling behavior, resistance to change if a psychological phenomenon of long standing interest in many applied branches of psychology.

Past travelling behavior has been acknowledged as a predictor of future action. Such as travelling behavior that is experienced as successful is likely to be repeated and may lead to habitual patterns. Some psychologists differentiate habit between two concepts, such as goal oriented and automatic oriented both. Although repeated past travelling behavior is addition goal oriented and automatic oriented. Further non-deliberative nature of habit may make appeals to judge and to predict future individual traveler's behavior accurately.

However, repeated one traveler will choose the destination to repeat to travel without a necessary constraint of goal orientation and automatic oriented both. So, it seems that psychological factor can influence any individual traveler why and how who choose to decide to repeat to choose the destination to travel.

So, survey investigation is only the traveler's thinking to answer the travel firm. It is not sure that the traveler's past travel experience is real answer. Otherwise, (AI) big data gathering method is computer gathering method which gather past traveler consumption actual data to analyze and conclude future traveler possible repeated travel destination choice and travel package choice more accurate.

The third reason is that on the strength of (AI) big data gathering method computer statistic view to predict future traveller consumer's destination and travel package choice. It is structural equation modeling is an extremely flexible linear-in-parameters multivariate statistical modeling technique. It has been used in modeling travel behavior and values since about 1980 year. It is a software method to handle a large number of variables, as well as unobserved variables specified as linear combinations ( weighted averages) of the observed variable.

1.3 Can (AI) big data gather data to predict when climate will change to influence poor travelling behaviours?

(AI) big data tool can predict the flexibility of human travelling behavioral change is at least the result of one such mechanism, our ability to travel mentally in time and entertain potential future. Understanding of the impacts is holidays, particularly those involving travel.

Using focus groups research to explores tourists' awareness of the impacts of travel own climate change, examines the extent to which climate change features in holiday travel decisions and identifies some of the barriers to the adoption of less carbon intensive tourism practices.

The findings suggest many tourists don't consider climate change when planning their holidays. The failure of tourists to engage with the climate change to impact of holidays, combined with significant barriers to behavioral change, presents a considerable challenge in the tourism industry. In the future, computer (AI) big data tool can attempt to predict when the country's climate change to influence travelers to choose to go to the country to travel, e.g. next month or next half year or next year hot travelling destinations.

Tourism is a highly energy intensive industry and has only recently attracted attention as an important contributions to climate change through greenhouse gas emissions. It has been estimated that tourism contributes 5% of global carbon dioxide emissions. There have been a number of potential changes proposed for reducing the impact of air travel on climate change. These include technological changes, market based changes and behavioral changes.

However, the role that climate change plays in the holiday and travel decisions of global tourists. How the global tourists of the impacts travel has on climate change to establish the extent to which climate change, considerations features in holiday travel decision making processes and to investigate the major barriers to global tourists adopting less carbon intensive travel practices.

It will bring this question: Will tourists aware the impacts that their holidays and travel have on climate changes to influence their travelling decision?

When, it comes to understand individual traveler's behavioral change, wide range of conceptual theories have been developed, utilizing various social, psychological, subjective and objective variables in order to model travel consumption behavior. These theories of travel behavioral change operate at a number of different levels, including the individual level, the interpersonal level and community level. Whether pro-environmental behavior can be used to predict travel consumption behavior in a climate change. However, the question of what determines pro-environmental behavior in such a complex one that it can not be visualized through one single framework or diagram.

Despite the potentially high risk scenario for the tourism industry and the global environment, the tourism and climate change ought have close relationship.

However, (AI) big data tool can be applied to find what factors to influence the time of travelers' travelling choices. What are the important factors and variables which can limit tourism? e.g. money, time, family problem, extreme hot or cold weather change, air ticket price, journey attraction etc. variable factors.

Mention of holidays and travel were deliberately avoided in the recruitment process, so as not to create a connection factor to influence traveler's individual mind. However, the dismissal of alternative transportation modes can be conceived as either a structural barrier, in the sense that flying is perhaps the only realistic option to reach long-haul holiday destination, or a perceived behavioral control barriers in that an individual perceives flying

as the only option open to whom.
The transportation tool factor will be depend to extent on the distance to the destination. This can also be interpreted in a social perspective as an intention with the resources available where much international tourism is structured around flying. To increase the availability of different transportation modes, tourists could choose holiday destination closer to home.

Finally, also how to predict future travel behavioral consumption. I feel that travel agents need to predict whether any country's random daily variation of weather factor is also important to influence travel behavior. e.g. in weather, temperature, rainfall and snowfall with traffic accidents factors will have relationship to cause travel demand.
Some scientists estimate suggest that when warmed temperatures and reduced snowfall are associated with a moderate decline in non-fatal accidents, they are also associated with a significant increase in fatal accidents. Thus increase in fatalities and temperature. Half of the estimated effect of temperature on fatalities is due to changes in the exposure to pedestrians, bicyclists and motorcyclists as temperature increase.
So, if any countries have rainfall, snowfall and low temperature to cause traffic accidents, whether this accident occurrence will influence the travelers who liking climb snow hills, riding bicycle, running sports who will avoid to travel to these countries' bad weather after occurs. So, why I feel that this natural climate factor will also be one serious factor to influence travel behavioral consumption. However, (AI) big data tool can predict more accurate than survey method when climate change to influence the country's climate to be poor, then it can predict when which countries are not popular acceptable to global country consumers' travel choice next month.

How can apply (AI) to provide travelling businesses with better-informed decisions ?

I shall explain how (AI) big data gathering technology can provide travelling businesses with better-informed decisions to drive top-line growth, deliver meaningful experience for travelling customers and smooth their path along the travelling consumer journey. The widely understood definition of (AI) involves the ability of machines or computers to learn human thinking, reasoning and decision-making abilities.
So, such as (AI) learning machine system can attempt to learn travelling

consumer's travel destination or travel package thinking, judgement of their reasons why they choose to go to the destination to travel or why they choose to buy the travel package and learn how and why they make their past travelling decisions from their past travel big data gathering.

A Narrative science study in 2015 year identified that (AI) was being used primarily in voice recognition, machine learning virtual assistants and decision support. This study also highlighted the many branches of (AI) and that techniques and their definition are used interchangeably. It is possible that (AI) can be used to gather big data , then to analyze to help travel businesses to predict travelling consumer travel destination and travel package choice behaviors. For example, one of the most common techniques is traveler machine learning, where algorithms are used to perform tasks by learning from the airline or travel agent whose past all travelers' travelling destination choice and travel package choice historical data.

However, during 2017 year, search engines will begin to find what additional factors can influence past traveler personal travelling destination and travelling package travelling behavioral data into prediction of future travelling customer behavioral results, such as the online traveler (user's) history of travelling data searches, such as anywhere are the most popular travelling locations or travelling destinations and previously captures conservations.

Artificial intelligence will use this past travelling destinations and travelling package information to power predictive search results, e.g. predictive future travelling consumer's choice behavioral processing for where will be their preferable travelling destination choice and how to design travelling package to satisfy future travelling clients' needs.

Predictive search will improve the quality of online travelling search results, and provide new insights into travelling consumers' travelling destination and package behavior and the moments which matter to them. Search will give recommendation into tailored how travelling consumer individual travelling destination choice in travelling decision making process. Several of the largest online platforms already use (AI) travelling machine learning to improve predictive travelling consumer behavioral search results.

For example, Google's rank brain technology adds research by understanding the context in which the travelling consumer has entered it. Over time, rank brain will learn further from user behaviors Amazon's DSSTNE ( pronouned destiny) learns from shoppers' purchasing habits and consumption behavior to offer better product recommend actions, which

Amazon can offer before a consumer has entered anything into the search bar.

Such as (AI) big data can gather past online travelers' e-ticket purchase transactions to conclude that online traveler's travelling choice habits and online traveler consumption behavior to offer better travelling destinations and travelling package opinions to travel agents or airlines. However, this technology is not independent of human input. For example, Google engineers will periodically retain the rank brain system to improve the models it uses.

For another example, in 2016 year , Apple computer revamped its travelling scene photos app to allow travelling consumers to search for specific travelling destinations in the travelling scene phots, they want to find anywhere travelling destination photos, not just dates and locations. Each travelling photo that an intelligent phone or intelligent pad user takes goes through 11 billion computations, so that travelling scene photos can understand exactly where is the travelling destination photography to let online travelling consumer to feel anywhere they plan to go to the location to travel. So, (AI) learning machine can make online travelling photos more attractive to influence potential travelers choose to the destination to travel after they see the travelling destination scene photos from internet.

It seems that in future, (AI) machine learning will allow online travelling search to evolve even further. Search engineers will deliver refined recommendations to airlines' online traveler e-ticket search users and use less human input to predict travelling consumers' needs from internet channel. For IBM computer example, it indicated 90% of the data that exists today has been created in the last two years.

This huge explosion of past traveler's e-ticket consumption data gives the opportunity to quickly spot and react to the latest trends, fashion and fads among its travelling clients and potential clients. This will allow airline or travel agent companies to better engage with younger travelling consumers, who gain influence access to the latest travelling destination and package trends.

They associate with to help define who they are as individuals. Thus, travelling company brands have to identify and make use of them before travelling consumers move on, but the vast quantity of past e-ticket purchase data available makes from internet channel. This a resource-intensive task. For next example, Lesara, a based online clothes store, uses this machine learning to inform its product decision often gathering information from internal and external sources.

When its trends -spotting shoes. Lesara has a range of over 20 styles and sells hundreds of pairs a day. It focus on giving consumers, the very latest trends allow Lesara to develop on average of 50,000 new items each year. It compared to 11,000 old items each year. Thus, travelling agents or airlines can attempt to apply (AI) big data gathering method to gather all past e-ticket purchase data, concerns where they prefer to choose to go to the destinations to travel and what travelling packages are the most attractive to the travelers to choose to buy. It aims to help them to predict where future travelers will prefer to choose to go to travel or what travelling package they will prefer to choose to buy next year.

For another (AI) big data prediction example, Lesara is one online clothes store, uses machine learning decisions after gathering information from internal and external sources. One of its most popular products, shoes with LED started life when its trend spotting software flagged up a blogger wearing similar shoes. Now Lesara has a range of over 20 styles and sells hundreds of pairs a day. Its focus on giving consumers the very latest trends allows Lesara to develop an average of 50,000 new items each year, compared to 11,000 for its competitor Lara.

It seems (AI) big data gathering machine learning can help Lesara business to predict what kinds of shoes design or style that shoe consumers will prefer choose to buy in future shoe market trend. Thus, Lesara can predict shoe consumers' taste successfully and it can manufacture many attractive style of shoes.

(AI) machine learning can gather global past shoe consumer's shoe shopping experiences, then analyzes to make conclusion to give lesara recommendation successfully. This will make the experience more enjoyable for shoe consumers and allow Lesara to advert whose different new style or design of shoes to deliver them move relevant messages by understanding the context of the experience.

So, online travel agents or online airline can also attempt to apply (AI) big data gathering method to predict where travelers will prefer to go to travel and how they ought design travelling packages to attract them to choose to buy next year. Hence, (AI) big data gathering technology can conclude how to design traveler agents' travelling package products to be the most attractive to excite many travelers choose to buy their travelling package, due to it has more accurate to predict travelling consumer destination and travelling package choice behaviors to compare human themselves prediction judgement effort, e.g. travelling survey or marketing research, or telephone enquire. It seems that (AI) machine judgement effort is more accurate to compare to human judgment effort in travelling industry.

## 2.1 Future travel consumption behavior

Can (AI) big data gathering tool predict traveler individual habitual behavior , e.g. renting travel transportation tools ?

Can (AI) big data gathering tool can predict past traveler destination and travelling package choice habit and it can be intended to predict of future traveler behavior to people are creatures of habits judgement of future anywhere travelling destination choice next year or next month or next half year destination prediction ?

Many of human's everyday goal-directed behaviors are performed in a habitual fashion, the transportation made and route one takes to work, one's choice of breakfast. Habits are formed when using the some behavior frequently and a similar consistency in a similar context for the some purpose whether the individual past travel consumption model will be caused a habit to whom. e.g. choosing whom travel agent to buy air ticket or traveling package; choosing the same or similar countries' destinations to go to travel ; choosing the business class or normal (general) class of quality airlines to catch planes.

Does habitual rent traveling car tools use not lead to more resistance to change of travel mode? It has been argued that past behavior is the best predictor of future behavior to travel consumption. If individual traveler's past consumption behavior was always reasoned, then frequency of prior travel consumption behavior should only have an indirect link to the individual traveler's behavior. It seems that renting travel car tools to use is a habit example. So, a strong rent traveling car tools useful habit makes traveling mode choice. People with a strong renting of traveling car tools of habit should have low motivation to attend to gather any information about public transportation in their choice of travelling country for individual or family or friends members during their traveling journeys.

Even when persuasive communication changes the traveler whose attitudes and intention, in the case of individual traveler or family travelers with a strong renting travel car tools habit. It is difficult to change whose travel behaviors to choose to catch public transportation in whose any trips in any countries. However, understanding of travel behavior and the reasons for choosing one mode of transportation over another. The arguments for rent traveling car tools to use, including convenience, speed, comfort and individual freedom and well known.

Increasingly, psychological factors include such as, perceptions, identity, social norms and habit are being used to understand travel mode choice. Whether how many travel consumers will choose to rent traveling car tools during their trips in any countries. It is difficult to estimate the numbers. As the average level of renting travel car tools of dependence or attitudes to certain travel package policies from travel agents. Instead different people must be treated in different ways because who are motivated in different ways and who are motivated by different travel package policies ways from travel agents.

In conclusion, the factors influence whose traveler's individual traveler destination choice behavior The factors include either who chooses to rent traveling car tools or who chooses to catch public transportation when who individual goes to travel in alone trip or family trip. It include influence mode choice factors, such as social psychology factor and marketing on segmentation factor both to influence whose transportation choice of behavior in whose trip. So, (AI) big data can be attempted to gather past traveler transportation tool choice, rent travelling car tools choice or catching public transportation tools choice to predict where destination can provide what kind of transportation tool to attract many travelers to choose to go to the place to travel.

How (AI) big data determine future travel behavior from past travel experience and perceptions of risk and safety for the benefits to travel consumers?

How (AI) big data determine future travel behavior from past travel experience and perceptions of risk and safety for the benefits to travel consumers? Why does individual traveler avoid certain destination(s) is(are) as relevant to tourist decision making as why who chooses to travel to others?

Perceptions of risk and safety and travel experience are likely to influence travel decisions. If travel agents had efforts to predict future travel behavior to guess whether travelers will feel where is(are) risk and unsafe to cause who does not choose to go to the country to travel. Then, the travel agents will avoid to choose to spend much time to design the different traveling package to attract their potential travel consumers to choose to travel. The reason is because in the case of individual traveler's tourism experience, the traveler whose past disappointment travel experience ( psychological risk) will be a serious threat to the traveler's health or life ( health, physical or terrorism risk). The past safety or unhealthy risk to the country(countries) will influence the traveler decides to choose not to go to the countries(country) to travel again in the future.

2.2 What is push and pull factors to influence any traveler who chooses where is whose preferable travelling destination ?

How to apply (AI) big data to predict individual traveler's behavioral intention of choosing a travel destination?

Understanding why people travel and what factors influence their behavioral intention of choosing a travel destination is beneficial to tourism planning and marketing. In general, an individual's choice of a travel destination into two forces.

The first force is the push factor that pushes an individual away from home and attempt to develop a general desire to go somewhere, without specifying where that may be.

The other force is the pull factor that pull an individual toward in destination, due to a region-specific or perceived attractiveness of a destination. The respective push and pull factors illustrate that people travel because who are pushed by whose internal motives and pulled by external forced of a destination. However, the decision making process leading to the choice of a travel destination is a very complex process.

For example, a Taiwanese traveler who might either choose new travel destination of Hong Kong or another old travel Asia destinations again or who also might choose any one of Western country, as a new travel destination. The travel agents can predict where who will have intention to choose to travel from whose past behavior and attitude, subjective and perceived behavioral control model. When (AI) big data gather past every country traveler number who chose to go to which countries to travel in order to judge where destinations will be the country travelers' travelling choice destinations in the future.

The factors influence where is the traveler choice, include personal safety, scenic beauty, cultural interest, climate changing, transportation tools, friendliness of local people, price of trip, trip package service in hotels and restaurants, quality and variety of food and shopping facilities and services etc. needs. So, whose factors will influence where is the individual travel's choice. It seems every traveler whose choice of travel process, will include past behavior. e.g. travelling experience, travelling habit, then to choose the best seasoned travelling action to satisfy whose travel needs. This process is the individual traveler's psychological choice process, who must need time to gather information to compare concerning of different travel packages, destination scene, climate change, transportation tools available to the destination, air ticket price etc. these factors, then to judge where is the best right destination to travel in the right time.

Hence, (AI) big data can gather past different countries' climate changing data, transportation tool changing data, destination scene environment changing etc. different data to give opinions to travelling businesses whether any country's these above factors will influence about how many traveler number will be increase or decrease in the future.

2.3 Why can expectation, motivation and attitude factor influence travelling behavior?

Social psychology is concerned with gaining insight into the psychological of socially relevant behaviors and the processes. For instance, on a global level bad influence to global warming, it influences some countries extreme cold or hot bad climate changing occurrence, then it ought influence some travelers' behavioral decision to change their mind to choose some countries to go to travel at the moment which do not occur extreme hot or cold climate ( temperature). e.g. above than 40 degree in summer or below than 0 degree in winter. Due to the extreme climate changing environment in the countries, it will cause them to feel uncomfortable to play during their trips. So, the global warming causes to climate changing factor will influence the numbers of travel consumption to be reduced possibly. This is global climate changing environment factor influences to bad or uncomfortable social psychological feeling to global travelers' mind of traveling decision. What is individual traveler expectation, motivation and attitude? Tourism sector includes inbound (domestic) tourism and outbound (overseas) tourism both incomes to any countries. According to recent article, a tourist behavior model has been developed, called the expectation, motivation and attitude ( EMA) model ( Hsu et al., 2010).

This model focuses on the pre-visit stage of tourists by modeling the behavioral process by incorporating expectation, motivation and attitude. Travel motivation is considered as an essential component of the behavioral process, which has been increasing attention from the travel; industry. The economic approach defines "tourism" is an identifiable nationally important industry. It includes the component activities of transportation, accommodation, recreation, food and related service. So, tourism behavioral consumption is concerned the individual tourist's usual habituate of the industry which responds to whose needs, and of the impacts that both the tourist and the tourism industry have on the socio-cultural, economic and physical environment.

However, travel motivation means how to understand and predict factors that influence travel decision making. According to Backman and others (1995, p.15), motivation is conceptually viewed as " a state of need, a condition that services as a driving force to display different kind of behavior toward certain types of activities, developing preferences, arriving at some expected satisfactory outcome." So, motivation and expectancy which has close relationship to any tourist before who decided to do any tourism of behavior.

Some economists confirmed motivation and expectancy which has relations, such as expectation of visiting an outbound destination has a direct effect on motivation to visit the destination; motivation has a direct effect on attitude toward visiting the destination; expectation of visiting the outbound destination has a direct affection on attitude toward visiting the destination and motivation has a mediating effect on the relationship in between expectation and attitude.

Hence, (AI) big data can gather all the country's climate environment change, transportation tool change, entertainment scene change, hotel price and restaurant price change etc. data to give opinions whether the country will attract how many traveler to choose to go to travel in the year.

What is (AI) deep learning techniques to forecast travelling environment behavioral consumption

Prediction how many travelers will choose to go to the country to travel. It is similar to apply deep-learning technology to predict how to raise the agricultural farming productivity in the agricultural export country.

The (AI) deep-learning technology leads to performance enhancement and generalization of artificial intelligent technology. It influences the global leader in the field of information technology has declared its intention to utilize the deep-learning technology to solve environmental problems, such as climate change.

So, it will help agriculture farming businesses can raise any plant food: vegetable, fruit, rice which grow up very easily if farmers can apply (AI)

deep-learning technology to solve environment problems to influence their plant food grow. If the whole year seasonal change is very good and it is suitable for any plant food to grow in farming land easily, e.g. rain is enough and soil is enough for any plant food to grow in the farm lands. Then, fruit, rice, vegetable etc. agriculture businesses will have much beneficial attribution to global farmers.

The question is how to use deep-learning technologies in the environmental field to predict the status of pro-environmental consumption. We predicted the pro-environmental consumption index based on Google search query data, using a recurrent neural network ( RNN model). To certify the accuracy of the index, we compared the prediction accuracy of the RNN model with that of the ordinary least square and artificial necessary network models.

For example, the RNN model predicts the pro-environmental consumption index better than any other model. we expect the RNN model to perform still better in a big data environment because the deep-learning technologies would be increasingly as the volume of data grows. So, deep-learning technologies could be useful in environmental forecasting to prevent damage caused by climate change to influence any rice, vegetable, tomato, potato, fruit etc. different plant food grow in any countries' farming land easily.

For South Korea example, over 800 government agencies spent 2.2 trillion Korea won on eco-products in 2014 year. However, green products are rarely purchased outside these agencies. This phenomenon occurs because there is a gap between consumer attitudes and behavior , that is environmental attitude is a major factor in decision making vis-a-vis the consumption of " green" food and services ( Jorea Ministry of Environment, 2015).

Therefore, it is necessary to understand those consumer attitude, that will lead to sustainability-conductive behavior and consumption. (AI) Deep learning system can be applied to attempt understand those traveler attitude to environment protection to fly to which country. For example, (AI) deep learning system can attempt to gather data concerns how many Hong Kong people concern air pollution challenge to influence their health, then it can attempt to predict how many Hong Kong travelers do not choose to go China travel, due to the air pollution challenge to influence their health.

Environmental travel consumption prediction

Recently, many researchers have studied pro-environmental consumption and household indexes as well as suicide rate predictions using messages posted by internet users on Google trend, Tweets etc. channel.

Whether can environmental consumption be predicted by (AI) deep-learning technological internet channel to influence how many travelers choose to go to the country to travel?

How can impact the pro-environmental consumption attitudes of green policies to influence how many travelers choose to go to the country to travel?

For example, Korea scientists estimated pro-environmental attitudes using search query data provided by Google trend and confirmed through regression analysis, that pro-environmental attitude has a positive correlation with the pro-environmental attitude index. They also explained that environment-friendly attitude of residents plan an important role in policy making. In the past, most household consumption indexed were calculated through surveys, but (AI) deep-learning technological tool " big data" have recently gained research attention ( Lee et al. 2016). So, (AI) deep learning technology can attempt to gather whether how many Korea residents who concern environment pollution to influence their eating green food attitude then to judge whether how many Korea residents hope to leave their country to travel anywhere either high risk environment pollution countries to travel or low risk environment pollution countries to travel in the future.

It seems that (AI) deep-learning technology can help agricultural export countries' farmers , e.g. US, UK, Canada, New Zealand, Australia, Japan, China, India etc. they can predict environmental behavioral consumption to any rice, tomato, potato , fruit, vegetable etc. plant food consumers. The beneficial advantages to them include as below:

(a) Assuming they know their countries' weather, when it has less rain to cause drought or when it has more rain in any seasonal time in the year. They can choose not to grow any kinds of above these plant food to avoid loss.

(b) They can make any kinds of above these plant food price raising after their prediction of these bad seasonal time to cause their plant food shortage supply challenge. Because these plant food consumers' demand number is more, but the supply of these above plant food supply number is less. However, due to they had predicted when the bad seasonal time can not allow them to grow these above plant food before. So, they have enough time to grow many these above plant food number in predictive good seasonal time to prepare to supply to their plant food import countries' plant food consumers to eat. Thus, these predictive environmental consumption plant food export countries can raise their plant food price to sell to them. When, the other non-pre-predictive environmental consumption plant food export countries can not supply any one of those plant food to them to eat, due to the bad climate to cause them can't grow any one of these plant food to export to sell.

Thus, (AI) deep-learning technology can be applied to predict how to raise the plant food supply number in order to raise price to the import plant food countries consumers to eat, due to they feel difficult to buy these plant food to eat in the bad climate seasonal time in whole year.

(c) (AI) deep-learning technology can help climate scientists to find what reasons cause their countries; rain sudden increases or cause their countries' rain sudden decreases. After its gathering data analysis, it can assist climate scientists to find solution methods to attempt to control the rain level can be right falling down level to let agricultural export farmers who can grow their plant food to sell to agricultural import countries in whole year.

(d) The agricultural export countries' farmers can apply (AI) deep-learning technology to help them to choose whether growing which kinds of plant food in that whether climate time to earn more plant food consumption number more easily.

Due to the agricultural countries climate will often change, for example, tomato, potato, rice, fruit etc. plant food can be adapt to grow in more rain time, but vegetable can not be adapt to grow in more rain time. If farmers can apply this technology to predict when it will have move rain or when it will have less rain to fall down in their countries. Then, they can choose to grow which kinds of plant food number more, in the suitable seasonal climate time in order to raise plant food growing number productivities to supply to sell to satisfy any agricultural food import countries' demand effectively.

(e) (AI) deep-learning technology can help agricultural import countries to solve agricultural food shortage challenge in long term. When this technology can be popular to base applied by the agricultural plant food export countries. It will solve global agricultural food shortage challenge. For example, when one agricultural export countries' farmers can popular accept to apply this technology to predict when to grow which kinds of plant food more to rise number productivities to sell. e.g. vegetable, fruit, rice Besides another agricultural export countries' farmers can also accept to apply this technology to predict when to grow plant food, e.g. potato, tomato to raise number productivities to sell. Then, they can concentrate on growing the specific kinds of plant food in order to raise the specific plant food number productivities in every seasonal change time every month. Then, global agricultural plant food supply must be raised, due to these predictive environmental change farmers can know who ought grow which kinds of plant food to sell to raise number productivities.

Consequently, (AI) deep learning can gather where countries will have high risk environment pollution to influence health food supply. Then, it can give opinions to travelling businesses when these high risk environment pollution countries will encounter the traveler number to be decreased, due to the environment pollution serious challenge will occur.

What methods can predict future travel behavioral consumption ?

How to use qualitative of travel behavioral method to predict future travel consumption from (AI) big data ?

I also suggest to use qualitative of travel behavioral method to predict future travel consumption. Methods such as focus groups interviews and participant observer techniques can be used with quantitative approaches on their own to fill the gaps left by quantitative techniques. These insights have contributed to the development of increasingly sophisticated models to forecast travel behavior and predict changes in behavior in response to change in the transportation system. I shall indicate the weaknesses of human travelling investigation methods as below:

First, survey methods restrict not only the question frame but the answer frame as well, anticipating the important issues and questions and the responses. However, these surveys methods are not well suited to exploratory areas of research where issues remain unidentified and the researched seek to answer the question "why?".

Second, data collection methods using traditional travel diaries or telephone recruitment can under represent certain segments of the population, particularly the older persons with little education, minorities and the poor. Before the survey, focus group for example can be used to identify what socio-demographic variables to include in the survey, how best to structure the diary, even what incentives will be most effective in increasing the response rate.

After the survey, focus, focus groups can be used to build explanations for

the survey results to identify the "why" of the results as well as the implications. One Asia Pacific survey research result was made by tourism market investigation before. It indicated the travel in Asia Pacific market in the past, had often been undertaken in large groups through leisure package sold in bulk, or in large organized business groups, future travelers will be in smaller groups or alone, and for a much wider range of reasons.
Significant new traveler segments, such as female business traveler. The small business traveler and the senior traveler, all of which have different aspirations and requirements from the travel experience.

Moreover, Asia tourism market will start to exist behaviors in the adoption of newer technologies, a giving the traveler new ways to manage the travel experience, creating new behaviors. This with provide new opportunities for travel providers. The use of mobile devices, smartphones, tablets etc. and social media are the obvious findings to become an integral part of the travel experience. Thus, quality method can attempt to predict Asia Pacific tourism market development in the future. It is such as (AI) big data gathering tool can give traveler quality opinions to any travelling businesses to make the more accurate where will be the popular travel destination choice next month or next half year or next year.

However, improving the predictive power of travel behavior models and to increase understanding travel behavior which lies in the use of panel data( repeated measures from the same individuals). Whereas, cross-sectional data only reveal inter-individual differences at one moment in time, panel data can reveal intra-individual changes over time. In effect, panel data are generally better suited to understand and predict ( changes in ) travel behavior. However, a substantial proportion was also observed to transition between very different activity/travel patterns over time, indicating that from one year to the next, many people renegotiated their activity/travel patterns.

How to apply advanced traveler information systems (ATIS) to predict future travelling behavior?

Nowadays, information can impact on traveler behavior and network performance. For example, when steadily growing levels of vehicle ownership and vehicle miles traveled information has been identified as a potential strategy towards man aging travel demand, optimizing transportation networks and better utilizing available capacity. Toward, this goal to predict further tourist behavioral consumption. Many countries, government tourism development institutes has applied advanced traveler information systems (ATIS) which travel behavior models and high-fidelity network performance models made increasingly feasible through the rapid advances in computer power. Crucial components of this problem domain are the modeling of individual tourist drivers' response to travel information and the development accurate guidance of relevance to real would trip makers. So, this advanced traveler information systems (ATIS) can assist the tourist who like to rent travelling car tools to travel in any countries own free traveler information systems service conveniently. Also, this travel information system can be intended to assist travelers to make better travel choices. e.g. this system can improve the decision making of individual traveler rather than improvements of network performance overall. So, we need to understand how tourists make their travel plans. Also, understanding decision process that lead to booking of the trip is equally important, as it allows of a potential behavior.

How can online tourism sale channel influence traveling consumption of behavior?

Nowadays, internet is popular, it seems that booking air ticket behavior of using internet is predicted to influence overall tourism air tickets payment method. Tourism industry has grown in the previous several decades. Despite its global impact, questions related to better understanding of tourists and whose habits. Using online travel air ticket booking benefits include booking electronic air tickets can be made from entering any electronic travel agents websites in the short time and electronic travel ticket payers do not need leave home, who can pay visa card to pre booking any electronic travel ticket from online channel conveniently.

How can analyze activity based travel demand ?

Nowadays, human are concerning the traffic congestion and air quality deterioration, the supply oriented focus of transportation planning has expanded to include how to manage travel demand within the available transportation supply. Consequently, there has been an increasing interest in travel demand management strategies, such as congestion pricing that attempts to change aggregate travel demand. The prediction aggregate level, long term travel demand to understanding disaggregate level ( i.e. individual levels ) behavioral responses to short term demand policies, such as ride sharing incentives, congestion pricing and employer based demand management schemes, alternate work schedules, telecommuting limitation of travel agent traditionally work nature shall influence oriented trip based travel modelling passenger travel demand indirectly.

Finally, online travel purchase will be popular to influence the number of travel behavioral consumption nowadays. Any travel package products can be sold from websites to attract travelers to choose to pre-book air ticket for any trips conveniently. In the past ten years, the internet has become the predominant carrier of all types of information and transactions. Regarding travel decisions, internet has also become an important sales channels for the travel industry, because it is associated with comparably lower distribution and sales costs, but also because it adapts to high supply and demand dynamics in this industry. Consequently, the travel and tourism industry tries to increase the internet sale specific share of sales volumes. So, internet sale channel has changed travel consumption behavioral pattern and characteristics and travel experience. For example, Switzerland has one of the highest population-to-computer ratio in Europe. It is also one of the most highly internet penetrated countries in terms of use of the WWW on a day-to-day basis, with more than 75 percent of the population older than 14 years using the WWW daily ( ICT, 2005).

The reason of booking online tourism may include: convenience, fast transaction, finding traveling package choice easily, more airline seats available. So, online booking tourism will influence the traditional tourism agents visiting of sales and air tickets and travelling package numbers to be decreased. Finally, the online booking tourism market shares will be expanded to more than traditional tourism agents visits sale market in the future one day. So, the travel agents who still use the traditional tourism visiting sale channel which ought raise whose features to compare to differ to online tourism sale channel if these traditional tourism agents want to keep competitive ability in tourism industry for long term.

What is actively based patterns of urban population of travel behavioral prediction method?

Actively based patterns of urban population. It is a method of motivational framework means in which societal constraints and inherent individual motivations interact to shape activity participation patterns. It can be used to predict one city or urban the numbers of travel demand in the year. It has two elements: First, capability constraints refer to constraints are imposed by biological needs, such as eating and sleeping and/or resources, such as income, availability of cars etc. to undertake the urban or city's family activities in the year. Second, coupling constraints define where, when and the duration of planning activities that are to be pursued with other individuals. So, this method needs to gather information ( data) to get the relationship between activities, travel and spending work time and space time to evaluate whether there are how many families who have real needs to spend time to go to travel in the year.

What is trip based versus activity based approaches?

What is trip based versus activity based approaches? The fundamental difference between the trip-based and activity based approaches is that the former approach directly focuses on trips without explicit recognition of the motivation or reason for the trips and travel. The activity based approach , on the other hand, views travel as a demand derived from the need to pursue travel activities. So, it is better understand the individual or family behavior basis for individual or family travelling decision regarding participation in travelling activities in certain places or cities or countries at given times and hence the resulting travel needs. This behavioral basis includes all the factors that influence the why, how, when and where of performed activities and resulting individuals and household, the cultural/social norms of the community and the travel surrounding environment.

Another difference between the two approaches is in the way travel is represented. The trip based approach represents travel as a collection of trips. Each trip is considered as independent of other trips, without considering the inter-relationship in the choice attributes , such as time, destination and mode of different trips. As tours are chains of trips beginning and ending at a same location , say home or work. The tour based representation helps maintain the consistency across and capture the interdependency and consistency of the modeled choice attributed among the trips of the same tour.

In addition to the tour based representation of travel, the activity based approach focuses on sequences or patterns of activity participation and travel behavior, using the whole day or longer periods of time is the unit of analysis. Such as approach can address travel demand management issues through an examination of how people modify their activity participation, for example, will individuals substitute more out-of-home activities for in home activities in the evening of who arrived early form work due-to a work schedule change?

The major difference between trip based and the activity based approaches is in the way, the time dimension of activities and travel is considered. In the trip based approach, time is reduced to being simply a cost making a trip and a day's viewed as a combination, defined peak and off peak time periods. On the other hand, activity based approach views individuals' activity travel patterns are a result of their time use decisions with a continuous time domain. As individuals have 24 hours in a day or multiples of 24 hours for longer periods of time and decide how to use that travel among or allocate that time to activities and travel and with who, subject to their socio-demographic, transportation system and other and scheduling of trips. So, determining the impact of travel demand management policies on time use behavior is an important step to assessing the impact of such policies on individual travel behavior. The final major difference between this two approaches relates to the level of aggregation. In the trip based approach, most aspect of travel, e.g. number of trips etc. are analyzed at an aggregate level.

Consequently, trip based methods accommodate the effect of socio-demographic attributes of households and individuals in a very limited fashion, which limits the activity of the method to evaluate travel impacts of long term socio-demographic characteristics of the individuals who actually make the activity travel choices and the travel service characteristics of the surrounding environment. So, the activity based models are better equipped to forecast the longer term changes in travel demand in response composition and the travel environment of urban areas. Also, using activity based models, the impact of policies can be assessed by predicting individual level behavioral responses instead of employing trip based statistical averages that are aggregated over defined demographic segments.

Can apply (AI) big data gathering method predict senior age will be main travelling target?

In the past, Germany government had established tourism survey analysis to analyze survey data in order to arrive at reliable conclusions on future trends in travel behavior. To aim to find how demographic change will influence the tourism market and how the industry can adapt to those changes. The travel analysis provided data on tourism consumer behavior, including attitudes, motives and intentions. Since, 1970 year, it is based on a random sample, representative for the population in private households aged 14 years or older. Then, a continuous high scientific standard combined with a national and international users makes the travel analysis a useful tool and reliable source for tourism industry and policy decisions. It aimed to gather statistical data. e.g. on the age structure and on demographic trends, quantitative and qualitative analysis with time series data from the travel analysis. It shows e.g. not only the future volume , quite different from today's seniors, or how who will travel of family holidays will change, e.g. single parents of low, but grandparents of growing significance for tourism.

Demographic change is said to be one of the important drivers for new trends in consumer traveling change behavior in most European countries ( e.g. Lind 2001). Because the growing number of senior citizens in the European Union and other industrialized countries, such as the USA and Japan, looks to become one of the major marketing challenges for the tourism industry. United Nations statistics predict that the share of people being 60 age or older will grow dramatically in the coming future, and is expected to rise from 10 percent of the world population in 2000 year to more than 20 percent in 2050 year ( United Nations Population Division, 2001). From its statistic, some data showed that travel propensity increased throughout life until the age of about 50 years of age and was then kept stable until very late in life 75 age. The most important results is that the travel propensity when getting older is not going down between 65 and 75 age of course, the overall development of this variable is influenced by a lot of other factors which are responsible for quite a variation over time. It is now possible to suggest that the general pattern of travel propensity is one of the key indicators for holiday life cycle travel behavior, includes three stages. The growth stage tends to increase from early adult hood until 45 age old or when reaching some 80%. The next stage is stabilization from the ages of around 50 age, until 75 age old, starting with a lower increase. Finally, the decrease stage is a slight decrease occurs once people reach the more advanced age of 75 age to 85 age old ( Lohmann & Danielsson 2001).

So, it seems Germany government tourism prediction to future travelers' behavior indicated these findings, such as on how future senior generations will travel, who had used survey data to examine the patterns of travel behavior of a generation getting older and applied the findings to draw conclusions on the future. Also, it predicted that on the future of family trips, family segmentation will be the travel behavior patterns in the future. These findings together with the statistical data on demographic change allowed for a better understanding of the coming tends in family holidays.

It's aim developed in consumer behavior related to demographic change and predicted what will happen future of tourism one had to consider other influences and drivers as well, for example, trends on the supply side. e.g. low cost airlines or in travelling consumption behavior in general whether how the past may provide a key to predict travel patterns of senior citizens to the future.

Given the projected growth of the senior citizens market, designing specific marketing strategies to meet the prospective needs of elderly tourists will become increasingly important. It has been an implicit assumption that it will be a close relationship between the travel behavior of today's senior citizens and the those of future ones. The growing number of senior citizens in the world. e.g. China, Hong Kong, Japan, USA etc. countries. Global senior citizen tourism market will be based solely on demographic predictions about the future of the population's age structure. However, many of these seniors won't only live longer but will be fitter and more active until later in life. Many of the will also have plenty in life. Many of them will also have plenty of time and money to spend on travel. So, will these new seniors behave like today's senior citizens? Will they adopt the same travel behavior as the previous generation or become a new market of oldies for the leisure and tourism industry? However, to determine the actual number of senior citizens who will be travelling and to sought to evaluate and specify certain difficult to predict the actual numbers of senior citizen to any country. However, they can be based on the implicit assumption that there is a close relationship between the travel behavior of past, present and future seniors. But is this a valid assumption? As the revise- analysis travel analysis survey, which was conducted in Germany every year, offered some interesting data possibilities. It was designed to monitor the holiday travel behavior, opinions and attitudes of Germans and has been carried out since 1970 year, questions in the questionnaire. Data are based on face to face interviews, with a representative sample of more than 7,500 respondents, the interviews being carried out in January each year. All results refer to the average for the defined generated, which ranges generally over ten years. The group of people then at the age of 60 to 69 age is described. This corresponds to the same generation ten years ago, when they had an age of 50 to 59 age. When this methodological approach is not necessarily very sophisticated, it does have the important advantages of being cost effective.

IS (AI) big data gathering method a better psychological method to compare human marketing research method predict travel behavioral consumption?

On the psychological view point, I think individual traveler's character will have those kind of personal characteristics. First, simplicity searchers value above everything ease not transparency in their travel planning and holiday making, and are willing to avoid having to go through extensive research. Second, cultural purists use their travel as an opportunity to immerse themselves in an unfamiliar looking to break themselves entirely from their home lives and engage. Sincerely with a different way of living. Third, social capital seekers understand that to be well travelled is a personal quality, and their choices are shaped by their desire to take maximum of social reward from their travel. They will exploit the potential of digital media to enrich and inform their experiences, and structure their adventures always keeping in mind they are being watched by online audiences. Finally, reward hunters seek a return on the investment who make in their busy , high-achieving lives. Linked in part to the growing trend of wellness, including both physical and mental self-improvement who seek truly extraordinary and often indulgent or luxurious' must have experiences.

Why needs to know the personal character of individual traveler's characteristics? Because if travel agents could feel which kinds of individual traveler's character, then who can predict which kind of travel package to design to them more easily. For example, how to determine future travel behavior from past travel experience and perceptions of risk and safety? We need to concern that the influences of past international travel experience, types of risk associated with international travel and the overall degree of safety feeling during international travel on individual's travelling experiences likelihood of travelling to various geographic regions on their next international vacation trip or avoidance of those regions, due to perceived risk. Because individual traveler's experience of safety risk degree to the countries, it will influence who chooses to go to the countries/country to travel again.

Why travelers avoid certain destinations are as relevant decision making as why who choose to go to the country(countries) to travel. Perceptions of risk and safety and travel experiences are likely to influence travel decisions; efforts to predict future travel behavior can benefit to individual tourist's decision making.

As Weber & Bottorn (1989) defined risky decision is as "choices among alternatives that can be described by probability distributions over possible outcomes" (p.114). Some psychologists judge subjective perceptions of physical reality, i.e. image of a particular tourist destination, whereas value judgement refers to the way individual rank destinations according to whose attributes. i.e. attractiveness, safety, risk etc. factors to form on overall image. So, if the individual traveler had unhappy and worried and unsafe experiences to go to where the place(country) to travel during whose vacation time before. Then, this negative travel experience will influence who is afraid to go to the place ( country) to travel again. Risk of place, country, destination or region means the danger is relatively high to the place, i.e. increasing in airplane accidents, crime or terrorist activity targeting citizens of potential traveler's nationality or the probability of occurrence is great , i.e. recent occurrences involving travel regions/ destinations under consideration or effective actions to control consequences exist. i.e. selecting safe regions and destinations, taking extra precautions when traveling to risky destinations. These risk factors will influence the individual traveler who chooses to cancel travel plan to go to the country again.

Another interesting research, how to predict behavioral intention of choosing a travel destination, which has focus of tourism research for years, but the complex decision making process leading to the choice of a travel destination has not been well researched. The planned behavior model using its core constructs, attitude, subjective norm and perceived behavioral control, with the addition of the past behavioral variable on behavioral intention of choosing a travel destination.

Understanding why people travel and what factors influence their behavioral intention of choosing a travel destination is beneficial to tourism planning and marketing. Understanding travel motivation is the push and pull model. The idea of the push and pull model is the decomposition of an individual's choice of a travel destination into two forces. The first force is the push factor that pushes an individual away home and attempts to develop a general desire to go somewhere else, without specifying where that may be. The second force is the pull factor, that pulls on individual toward a destination, due to a region specific travel location or perceived attractiveness of a destination. The respective push and pull factors illustrate that people travel because who are pushed by their internal motives and pulled by external forces of a destination. Nevertheless, how push and pull factors guide people's attitude and how these attributes lead to behavioral intentions of choosing a travel destination have rarely been investigated. The decision making process leading to the choice of a travel destination is a very complex process. The planned behavior model is as a research framework to predict the behavioral intention of choosing a travel destination. The model based on the three constructs of attitude, subjective norm, and perceived behavioral control ( Fishbein & Ajzen, 1975).

In conclusion, the factors can influence travelers who decide to choose to travel the country, which include personal safety was perceived to the highest motivation factors among the important factors which include, scenic beauty, cultural interests, friendliness of local people, price of trip, services in hotels and restaurants, quality and variety of food and shopping facilities and services. The factors include both push and pull. Push factors include knowledge, prestige, and enhancement of human relationship etc., whereas, the most significant pull factors include high technologic image, expenditure and accessibility etc. For example, Japanese travelers visiting Hong Kong. Push factors are such as exploration dream fulfillment and pull factors are such as benefits sought, attractions and good climate city. It will be the factor of future travel patterns and motivations of sub-cultural and ethic groups for Japanese choice to go to Hong Kong travelling.

How can apply (AI) digital channel ( big data gathering method) predict travelling consumer behaviors?

(AI) big data digital channel can be applied to help travelling businesses to evaluate whether how much the e-ticket price and travelling package price is the most attractive or reasonable to persuade travelling consumers feel it is the most reasonable price to choose to buy the airline's e-tickets or the travel agent's travelling package product from internet channel . It helps travelling consumers to feel which airlines or travelling agents which ought change their e-ticket and/or travelling package price to let travelling consumers to choose to buy the airline e-ticket or the travelling agent's travelling package products from internet channel. It can be applied to predict whether how many travelling consumer numbers can be increased or decreased when the airline e-ticket price is variable or the travelling agent travelling package price is variable . It aims to give opinions to help any online airlines or travelling agents to judge whether which e-ticket or travelling package price is the most reasonable to let travelling consumers to accept to choose to buy which airline's e-tickets or traveling agent's package products more attractive.

Thus, (AI) e-ticket or e-travelling package price measurement technology can be preference to be applied online communication ecommerce and mobile phone internet platform aspect. As traveling businesses can enter their past e-ticket or travelling package prices data and past travelling customer number data into computer or mobile. Then, (AI) price measurement technology can gather these data to analyze these e-ticket or travelling package product prices and past travelling customer number to compare their e-ticket and/or travelling package prices variable changing range level to find their e-ticket and /or travelling package price variable difference to measure to make conclusion about every travelling package or/and e-ticket product's price variable changing will influence how many travelling customer number increase or decrease changing to choose to sell their different kinds of travelling package or e-ticket products more accurate. Then, (AI) price measurement software will help them to analyze all past e-ticket and/or travelling package price variable changing data to compare whether which e-ticket and/or travelling package price range can let travelling customers to feel it is more reasonable and attractive to influence them to choose to buy their e-ticket or travelling package product among different airlines and travel agent choices. Because any e-ticket or travelling package product's price is one important factor to influence travelling consumers to choose to buy the airline's e-tickets or travelling agent's travelling package products.

For example, Amazon publish has applied (AI) price measurement technology to help authors to decide how much every different topic of e-book or paper book price, it can attract the largest number of readers to buy. Any one author only needs to type whose book name to Amazon publish author himself/herself Amazon website. Amazon publish (AI) price measurement learning machine will help them to auto-calculate and judge how much e-book or paper book price is the most attractive and the most reasonable in order to increase reader number to buy their e-books or paper books to read. So, (AI) online price measurement machine will gather past similar book names and past every similar book readers' reading times and the number of readers to give opinions to let every author to judge whether his/her very new e-book or paper book ought charge how much price to the e-book or paper book which can attract many readers to choose to buy. Although, it is not ensure that the e-book or paper book price must let readers to feel it is the most reasonable price to choose to buy in reader's view point. However, it has other factors to influence readers' choice to buy the e-book or paper book, e.g. whether the book content is attractive to public, the author's familiarity, the book's page is enough or not to satisfy readers to read etc. factors. But, instead of all these extra factors to influence readers to choose to buy the book to read. (AI) price measurement learning machine can real give opinions to every author to let them to judge the e-book or paper book different price range whether is too high to influence readers to choose to buy to read or tool low to influence readers feel it is possible poor content book to compare other similar content books. Thus, (AI) price measurement machine can help authors to predict every reader's reading behaviors or reading experience and reading habit from online channel in short time easily. The author only enter the book name to let Amazon publish price measurement machine to check, it will follow past reader's reading habit and reading experience to judge whether the similar all book topic sale record to judge how much price is the reasonable price to attract many readers to buy the book.

Hence, (AI) can be applied to digital channel to help travelling businesses to predict travelling consumer behavior in the future. In the future, mobile/ smartphone, laptop, desktop will be most frequent used ecommerce channels to develop online business. So, (AI) can be also applied to these platforms to gather data to make analysis to help travelling businesses to predict travelling consumer purchase behaviors popularly. Due to , ecommerce is popular to global, so digital online and instore channels can be one good channel to let (AI) learning machine to make platform to gather past every online travelling consumer purchase ( buying) experience data to help travelling businesses to build airline or travelling agent brand personality and having a responsible, positive impact on society.

To apply (AI) learning machine technology to understand travelling customer online purchase behavior, it will raise business e-commerce successful chance: For example, (AI) learning machine can help travelling businesses to gather data to analyze to determine whether short-term or long-term signals in the online travelling consumer behavior that indicate higher purchase intents to let every online travelling business to know. (AI) learning machine can find that online users with long-term purchasing intent tend to save and click through on more content.

However, as online travelling users approach the time of purchase their activity becomes more topically focused and actions shift from saves to searches from online travelling consumption channel. Then, (AI) learning machine will further find that the brand airline and/or travelling agent purchase signals in online travelling consumption behavior can exist weakness before an online travelling purchase is made and can also be traced across different online travelling purchase categories. Finally, (AI) learning machine synthesize these insights in predictive models of online travelling user purchasing intent to the brand of airline or/and travelling agent travelling package product. Taken together, it's work identifies a set of general principles and signals that can be used to model online travelling user e-ticket and/or travelling package purchasing intent across many online content discovery applications. Thus, (AI) learning machine can help online travelling businesses to gather any online travelling users' click online travelling behaviors data to judge whether there are how many online travelling users will choose to find their online travelling business websites to make final decisions to buy their travelling package or/and e-ticket products from online channels. Then, it will give opinions to help the online travelling businesses to let it to judge whether what are the important website factors will help its online travelling business to attract many online travelling consumers, e.g. designing unattractive travelling website issue, online unattractive scene photos issue, unclear website travelling photo color issue, unclear website travelling advertisement message, contents and words impressions issue, lacking image movement frequent attractive seeing issue etc. different website factors. Thus, online digital channel will be one good choice to apply (AI) learning machine to help travelling businesses to predict travelling consumer behaviors.

Thus, (AI) big data technology can also assist travelling consumers to gather different manufacturers' data to compare what their advantages and disadvantages of their travelling package products are. Then, travelling consumers can make comparison to choose which airline or travelling agent is the suitable to whom to buy e-ticket or pre-booking travelling package in online travelling consumption market.

.

Thus, I believe that artificial intelligent "big data" gathering method can be suggested to be applied to attempt to predict travelling consumer behavioral changes in global online travelling business environment, the reasons are as below:

On the travelling consumer's beneficial hand, travelling consumers can apply this (AI) big data gathering method to attempt to gather any global airline e-tickets and/or travelling agent's package product data to be analyzed by this artificial intelligent learning system to compare human general marketing research method, e.g. survey, questionnaire, marketing plan etc. different human judgement methods to predict traveler consumption behavioral change model. Then, it analyzed all the different data to compare what are the range of the most reasonable e-ticket and/or travelling package online purchase history and sale in order to make more accurate prediction to future traveler change traveling consumption behavioral model in next month, or next half year or next year short term period traveling consumption change prediction.

Reference

Backman and others "motivation is conceptually viewed as " a state of need, a condition that services as a driving force to display different kind of behavior toward certain types of activities, developing preferences, arriving at some expected satisfactory outcome.", 1995, p.15.

Fishbein & Ajzen, "The model based on the three constructs of attitude, subjective norm, and perceived behavioral control". 1975.

Hsu et al. "A tourist behavior model has been developed, called the expectation, motivation and attitude " ( EMA) model ,2010.

ICT,WWW . "Switzerland has one of the highest population-to-computer ratio in Europe." Switzerland, 2005.

Jorea Ministry of Environment, " For South Korea environmental attitude is a major factor in decision making vis-a-vis the consumption of " green" food and services", Korea, 2015.

Korea Ministry Of Environment. Public Organizations spend 2.2 Trillon Korean Won To

Purchase green Products in 2014; Ministry Of Environment: Sejoung, Korea, 2015.

Lind , Lohmann & Danielsson , United Nations Population Division, "Demographic change is said to be one of the important drivers for new trends in consumer traveling change behavior in most European countries". 2001.

Mayne, Lonnie. " Evolve of die in the age of the consumer". Entrepreneur, N.P. , 16 Apr. 2014. web of Oct. 2016.

Lee, D.; Kim, M. ; Lee, J. adoption of green electricity policies: Investigating the role of environmental attitudes via big data-driven search-queries. Energy policy 2016. 90, 187-201.

Lee, Terrence, " Tech in Asia-connecting Asia's startup system " Tech. in Asia- connecting Asia's startup ecosystem, N.p.,4 July 2016.

Weber & Bottorn "risky decision is as choices among alternatives that can be described by probability distributions over possible outcomes" , 1989, p.114

Bibliography

Adrian, P. (2012). Introduction to marketing theory & practice, 3 rd edition, London: Oxford press.

Ajzen, I (1991). The theory of planned behavior. Organizational behavior and human decision processes, 50(2), 179-211. doi: 10.1016/0749.5978 (91) 90020-7.

Alba, Joseph W. and J. Wesley Hutchinson (1987). " Dimensions Of Consumer Expertise", Journal of consumer research, 13 March, 411-454.

Bailey, L., Mokhtarian, P.L. Little, A. (2008). The broader Connection Between Public Transportation, Energy Conservation And Greenhouse Gas Reduction, Report Prepared As Part Of TCRP Project J-11/Tasks Transit Cooperative Research Program, Transportation Research Board Submitted To American Public Transportation Association in http://www.apta.com/research/into/online/land_use.cfmi, accessed 17 April 2008.

Baucer, R,"Consumer Bhavior As Risk Taking , In Risk Taking And Information handling In Consumer Behavior", D. Coxceds Harvard University Press, Cambridge, Mass 1976.

Biederman, P. (2008). Travel and tourism, Pearson Prentice Hall, New Jersey.

Bogers, R. P., Brug, J. Van Assema, P., & Dagnetie, P.C. (2004) , Explaining fruit and vegetable consumption: The theory of planned behavior and misconception of personal intake level. Appetite, 42,157-166.

Bolton, Ruth N. (1998), " A Dynamic Model Of The Duration Of The Customer's Relationship With A Continuous Service Provider: The Role Of Satisfaction", Marketing Science, 17 (1), 45-65.

B.Shiv and A. Fedorikhin, " Heart And Min In Conflict: The Interplay Of affect And Cognition In Consumer Decision Making", J. Consumer Res., vol. 26, pp. 278-292, Dec. 1999.

Brown, K.W., Ryan, R.M. Reswell , J.D. (2007). Mindfulness: Theoretical Foundatins And Evidence For Its Salutary Effects. Psychological Inquiry, 18, 211-237.

Burke, R.R. : Behavioral effects of digital signage, J. Advertising Res. 49(2), 180-185 (2009).

Cant, M., Brink , A. & Brijall, S., Consumer behavior, Cape Town, South Africa: Juta, 2006.

Conner, M. & Abraham, C. (2001). Conscientiousness and the theory of planned behavior: Toward a more complete model of the antecedents of intention and behavior. Social psychology bulletin, 27, 1547-1561.

Cooper C. Mallon, K, Leadbetter S, Pollack L, Peipins ( 2005) , cancer internet search activity on a major search engine, United States 2001 to 2003, J Med Internet Res. 7(3): e36.

Cope, R. R. Cope and H. Davis (2008). Disney's virtual Queues: A strategic opportunity to co-brand services ? Journal of Business & economics research, vol. 6 no10, 13-20.

Cornelia, B.F. (1999) Rural development news, the North Central Regional Center For Rural Development vol. no 24 , IOWA.

Couper, M.P. J. Blair and T. Triplet ( 1999). A Comparison Of Mail And E-mail For a Survey Of Employees In USA Statistical Agencies. Journal Of Official Statistics, 15, 39-56.

David J. Nowak & Gordon M. Melsler (2016) " Air quality effects of urban trees and parks." National recreation and park association, USA.

Data monitor ( 2008). The proctor and gamble company. Retrieved Nov. 15 2009 from http://www.datamonitor.com/

De Hollander, A. E. M., J.M. Melse, Elebret & P. G.N. Kramers ( 1999), " An Aggregate public health indicator to represent the impact of multiple environmental exposures" Epidemiology: 606-617.

De Visser, R.O., & McDonnell, E.J. ( 2013). " Man points": Masculine capital and young men's health. Health psychology, 32( 1), 5-14. doi:10. 1037/ a0029045.

Dunn, J & A Neumsister (2002). Knowledge management in the Information age. E. business review, Fall , 37-45. Jounral of service, spring 2011, vol. 4, no1, De Grovte (2009).

Dyer, D., F. Dalzell & R. Olegario ( 2004). Rising tide. Lessons learned from 165 years of brand building at Procter and Gamble. Boston, MA: Havard Business School Press.

Eysenbach G (2006) Infodemiology: Tracking flu- related searches on the web for syndromic surveillance. American Medical Informatics Associaion Annual Symposium Proceedings , Curran Associates, Red Hook, NY, pp. 244-248.

Ettredge M, Gerdes, J. Karuga , G (2005) Using web- based search data to predict macro-economic statistics. Commun ACM 48: 87-92.

Felce, D. and Perry, J. (1995). Quality of life: A contribution to its definition and measurement, vol. 16, no.1 pp: 51-74.

Feldman, Jack M. And John G. Lynch Jr. (1988), "Self-
Generated Validity And Other Effects Of Measurement On Belife, Attitude, Intention And Behavior", Journal of applied psychology, 73(3),421-35.

Fiese, M, Hofmann, W., & Wanke, M (2009). The impulsive consumer. Predicting consumer behavior with implicit reaction time measurement. In M. Wanke (ed.) Social psychology of consumer behavior (pp.335-364). New York, NY: Psychology press.

Fitzsimons, Gavan, J. And Vicki G. Morwitz ( 1996), " The Effect Of Measuring Intent On Brand-Level
Purchase Behavior", Journal of consumer research, 23 (1), 1-11.

Hallerman , D. (2008) video Advertising Online: Spending And Pricing , New York. E-Marketer.

Harriet Griffey. (2010) The art of concentration, enhance focus, Reduce, stress and achieve move. Macmillan publishers ltd,Basinastoke and Oxford, London UK.

Helleman, D. (2008) Video Advertising Online: Spending And Pricing , New York, E-Marketer.

Hensen, C. (2003). Kreuzfahrtourismus.www.christoph- hensen.de/ Facharbeit.pdf.

Huang, H.I. (2012). An empirical analysis of the strategic Management of competitive advantage: a case study of higher technical and vocational education in Taiwan ( Doctoral dissertation,
Victoria University).

Jamieson, Linda F. And Frank M. Bass ( 1989), " Adjusting Stated Intention Measures To Predict Trial Purchase Of New Products: A Comparison Of

Models And Methods," Journal of marketing research, 26 ( August), 336-45.

Korea Ministry Of Environment. Public Organizations spend 2.2 Trillon Korean Won To Purchase green Products in 2014; Ministry Of Environment: Sejoung, Korea, 2015.

Kremers, S.P. J., De Bruijn, G.J., droomers, M., Van Lenthe, F. J., & Brug, J. (2005). Environmental interventions for selected dietary behaviors in adults. In J. Brug & F. J. Van Lenthe ( eds.) , Environmental determinants and interventions for physical activity, nutrition and smoking: A review pp. 282-315. Rotterdam: Erasmus Medical Center.

Lee, D.; Kim, M. ; Lee, J. adoption of green electricity policies: Investigating the role of environmental attitudes via big data-driven search-queries. Energy policy 2016. 90, 187-201.

Lee, Terrence, " Tech in Asia-connecting Asia's startup system " Tech. in Asia-connecting Asia's startup ecosystem, N.p.,4 July 2016.

Los Angeles Country Department Of public Health (2016), Country Health Ranking Model, Retrieved From www.countryhealthrankgings.org/our-approach. USA.

Mayne, Lonnie. " Evolve of die in the age of the consumer". Entrepreneur, N.P. , 16 Apr. 2014. web of Oct. 2016.

McGregor, S.L. T., & Goldsmith, E.B. (1998). Expanding our understanding of quality of life, standard of living and well-being. Journal of family and consumer science, 90(2), 2-6, 22.

McMichael, A.J. M. Mckee, J. Shkolnikov and T. Valkanen ( 2004), " Morality trends and setbacks, global convergence or divergence?", Lancet 363, 1155-1159.

Melse, J.M. & A.E. M. De Hollander (2001). " Human Health And The Environment", background document for the OECD Environmental Outlook, OECD, Paris.

Moschis, George p. & Roy, L. Moore ( 1979), " Decision making among the young. A socialization perspective " Journal of consumer research , 6 ( September).

Mulligan, M. Banerjee, T & Thomas, N. (2008) ,European Paid Content And Activity Forecast, (2008 to 2013), Jupiter Research.

Peter, J., Ryan, M, M, " An Investigation Of Perceived Risk At The Brand Level, " Journal of marketing research, 13 May 1976, pp. 184-188.

Pieters, R., & Wedel, M. (2007). Goal Control Of Visual Attention To Advertising: The Yarbus Implication. Journal Of Consumer Research, 34, 224-233 ( August).

Parasuaman, and Leonard L. Berry (1985), " Problems And Strategies In Sevices Marketing", Journal of marketing, 49 ( Spring), 33-46.

Priesnitz, W. (2007) Counting Our Food Miles. Natural Life, 1 July.

R.C. Oliver, " When is consumer loyalty?" J.Marketing vol. 63, pp.33-44.1999.

Reggiani, A . (ed). 1998, accessibility, trade and locational behavior, Ashgate publishing ltd, England.

Rushe, D. (2013) " The 10 best paid CEO in America". The Guardian , 22 Oct, ( online). Available at: http://www.theguardian.com/business/2013/Oct22/best-paid-chief-executives-america (Accessed: 3 May 2014).

Spiekermann and Wegener (2007), update of selected potential accessibility indicators. Final report, urban and regional research ( S&W), RRG spatial planning and geoinformation. ESPON. Available online at http:// <www.espon.eu/mmp/online/website/ contentprojects/947/1297/ file_2724/espon_accessibility_update-2006-fr_070207.pdf>, accessed on 1 July 2009.

Starbucks (2014) Our company available at http:// www. starbucks.com/ about- us/company-information ( accessed: 3 May 2014).

Shostack, G. Lynn ( 1984), " Designing Services That Deliver", Harvard Business Review, 62 ( January-February), 133-9.

Shostack, G. Lynn (1985), " Planning The Service Encounter ,in the service encounter" , John A. Czepiel, Michael R. Solomon, and Carol F. Suprenant, eds. New York: Lexington Books, 243-54.

Shostack, G. Lynn (1987), " Service Positioning Through, Structural Change", Journal of marketing, 51 ( Janurary), 34-43.

Soloman, Michael R. (1985), "Packaging The Service Provider", Service Industries Journal , 5(1), 64-71.

Stevens, C.W. (1980), "K-MartStores Try New Look To Invite More Spending" The Wall Street Journal, Nov. 26, 29-35.

Sullivan, Nicholas P(2007). You can hear me now: How Micro loans and cell phones are connecting the world, San Francisco, CA: John Wilsey & Sans, 2007.

T. Ambler, A. Ioannides, And S. Rose, " Brand s On The Brain : Neuroimages Of Advertising ", Business Strategy rev., vol. 11, 3. pp. 17-30. 2000.

Westbrook, Robert A. ( 1980), " Intrapersonal affective influences on consumer satisfaction with products, " Journal of consumer research , 7 ( June) 49-54.

Wiig, k.(1993). Knowledge management foundations: Thinking About thinking. How people and organizations create, represent and use knowledge vol.1 , of knowledge management series schema press: Arlington, TX.

World Health Organization (2003). Diet, nutrition and the prevention of Chronic diseases report of a joint WHO/FAO. expert consultation. Geneva: World Health Organization.

Wysocki, B. (1979), " Sight, Smell, Sound: They're all arms in retailer's arsenal" The Wall Street Journal, Nov. 17, 1979. 1-35.

Yale Center For Environmental Law And Policy (2006). Environmental Performance Index. Data available on-line at http://epi.yale.edu

Can artificial intelligence development to tourism industry

● Artificial intelligence will bring what benefits to
Influence traveler consumption behavior or raising
Travelling leisure desire?

Artificial intelligence will bring what benefits to influence traveler consumption behaviors or raising travelling leisure desire? It would be unreasonable not to analyze how artificial intelligence is affecting, such a big part of global economy. Especially, how it leads travel media to develop. We need to answer those questions , before we hope predict how artificial intelligence can bring advantages to influence travelers to raise travelers travelling leisure desires in long time. Which tends will shape the travel industry? What will customers demand in their journeys? Which business models will prevail ? What role of artificial intelligence to play in future tourism industry? What new opportunities and business models do the new technology block chain and artificial intelligence offer for the travel industry? How is artificial intelligence future impacting the travel industry ? The utilization of artificial intelligence will how raise tourism demand?

Future artificial intelligence tourism development may help travel leisure benefits on these several aspects. They may include geographical elements choice, the tourist generating regions choice, the tourist destination regional choices and the transit route regions choice.

It is one exciting tourism service arrangement for travelers when artificial intelligence is participated to any travel agents' travelling leisure provision services. Artificial intelligence can help any travelling customers to arrange whole journey of a tourists, in these stages . In the before booking stage, (AI) journey service arrangement may include these from searching and being inspired stage to discovering and planning stage and booking stages. During booking, (AI) journey service arrangement may include booking and defining and improving stages. Finally, in the alter booking stage, (AI) journey arrangement service may include experiencing and reflecting and share , it is the tourist's leisure feeling after he/she had experienced the whole travel journey service arrangement. So, the (AI) 's whole tourism journey arrangement whether it is suitable to the tourist to let him/her to feel satisfactory ; or more satisfactory ; or the most satisfactory enjoyment feeling to his/her journey. It depends on how the (AI) judges the whole tourism journey service arrangement, e.g. reasonable price and comfortable accommodation choice for the business traveler of leisure travel in order to satisfy his/her accommodation choice. Because artificial intelligence is one complex computer machine, it ought may help any tourists to search accommodation information to search the most reasonable price and the most comfortable accommodation choice from social media to compare the tourist himself/herself accommodation search activity in order to get the most accurate accommodation choice and in order to satisfy the tourist's living need in his/her whole journey. So, future artificial intelligence is applied to tourism leisure service arrangement aspect. (AI) is this a goal oriented execution for any one tourist before he/she makes the final accommodation choice, tourism destination choice. It needs to help the tourist to execute his/her whole journey process, no matter how difficult this process may seem for humans.

- How (AI)helps tourism industry to reduce cost ?

In the future, robotics or artificial intelligence systems dominate the tourism industry, they may include chat bots, travel assistants, and service robots to any travel agents service providers. Their roles may include these several aspects to any travel agents service providers. They may include searching and being inspired, discovering , planning and booking , refining and improving, how excites the tourist experiencing to the journey service arrangement, reflecting and how raising his/her tourism leisure enjoyment feeling during the whole tourism journey from customer intelligence platform service.

However, future travel agent intermediaries as well as inventory providers will start to focus on their quality of offering and journey arrangement service. Thus, they invest comparable less in operational excellence, e.g. concentration of digital systems at the search and booking phase and robots at the experience phase of the customer journey is almost self-explanatory, journey searching, planning and booking via internet is nowadays the norm. This, these are no hybrid systems necessary. However, during the experience phase, in which the traveler leaves the realms of the digital world, physical interaction, which can only be delivered by robots, regains importance. Since char bots , travel assistants as well as service robots are all representatives of systems used at the travelling customer interface. However, supposing when the travelling agent decides to install (AI) systems, can not be explained by the trend of a rebound on customer journey experience. However, during the expert discussions, it was revealed that most of the tourism industry does not consider technology a core competency and often lack the capacities to self-develop and install such (Ａ Ｉ ) systems.

However, it may also seem that AI can impact automatization tasks to tourism industry, chat bots or service robots are typical representatives in which artificial intelligence is used to automate a task previously done by humans. This can be called the automatization effect. Nevertheless, automatization through (AI) systems should not be equated with other forms of technology utilization which are called automatization too. Nowadays, the automatization through self-serving terminals, whether at airports through self-check terminals or at hotels through self check in systems is quite common in the tourism service industry. However, in travel agents service industry sector, the (AI) systems process itself is not automated, but rather changed from the airport check in / out organizations to the travel agent journey arrangement customer service organizations. Artificial intelligence on the other side truly automates these process. So that neither a human at the airport or travel agent journey service arrangement organizations nor the travelling customer much do it.

A recent example are (AI) supermarkets , such as the “ Amazon Go” store in which can artificial intelligence system, which can recognize persons and objects, tracks people and the objects which they put in their shopping-bad and automatically bills their purchases on the customers bank account (Ｓｔａｒｋ，２０１７)·

Another similar system is imaginable at hotels and airports in which a face-recognition system recognizes the arrival of guests and automatically checks them in. JetBlue is already testing such a system a Boston airport (Entis, 2017).

However, how tourism organizations apply (AI) system can be employed to increase its competitive advantages, it means that either be achieved through differentiation, meaning delivery supervisor benefits to the consumer, or lower cost. If the robotic can help any tourists to choose the cheapest and the most comfortable apartment to let the tourist to live in his journey, such as recommender systems in the booking stage generate further benefits for consumers. The effect, however, is often related to automatization cost deduction, such as(AI)system can help the tourist whose the whole journey apartment cost reduction. Thus, one could assume that (AI)systems which automate tasks are mainly implemented to reduce apartments living cost for the tourist's whole journey. For example, chat bots can undoubtedly be used to replace sales and service staff at the tourist interface.

However, many companies use chat bots to expand their sales and service offerings instead, without reducing the respective staff. AirFrance-KCM uses its chat bots to sell its products and inform its customers about flight schedules via new channels, such as the face book messenger. This service is available 24 hours a day and seven days a week. This is mainly because chat bots are pure digital systems and can this practically scaled –up indefinitely ( Kelly , 2016).

A simple example is that many hotels utilize robots as a marketing tool to attract guests than as a true mean to save costs, it nevertheless shows the feasibility of such as model. A simpler use case would be the utilization of a vacuum cleaning robot to automate the task of cleaning the floor. It would replace a share of the existing cleaning personal and this gave costs. However, also (AI) systems which improve an existing task or just enable one, can and are used for cost savings. For instance, the British Airline Easyjet uses (AI) systems to better predict the demand of food and beverage on its flights. It reduced the inventory costs for the airline and help it to sustain a cost advantage ( Emarkerer , 2017).

Reference

Emarketer 2017, " How EasyJet uses artificial intelligence to improve operations-eMarketer" Emarketer. Accessed Dec. 10
http://www.emarketer, com/article/how-easyjet-uses.

Artificial-intelligence-improve-operations/1014558.
Entris, Laura: 2017 " Delta, Jetblue flights to test facial recognition scaual Fortune." Fortune.com accessed Dec. 2019. http://fortune.com/2017/06/01/jetblue-delta-boarding-checkin/
Stark, Hareld. 2017 " Amazon go, A cashierless convenience store Now available near you' Forbes.com accessed Dec. 9
http://www, forbes. Com/sites/haroldstark/2017/11/20 introducing-amazon-go-a-cashierless-convenience-store-available-near you/20efofo952f9.

Hence, future robotic may be applied to apartment service industry in hotels, instead of travel agents or airport services. For example, vending machines are found in public locations are doing the work of a hospitality workers and ATM machines have largely replaced the human labor of the bank teller. The design of robot-friendly hospitality facilities may bring cost benefit of adoption of robots for travel, tourism and hospitality companies. They are service robots.

In tourism service industry, customer attitudes can be influences in the mind of the tourist and is considered as an important element to influence the hotel, travel agent, airport services success. So service robotic participation will influence service robotic can bring positive attitude to their customers, such as providing excellent service for hotel book room service, air ticket check in /out service for airport travelers or apartment choice service for tourist journey arrangement choice services.

Thus, (AI)service will influence the country's overall hotel room booking, travel leisure and airport check in arrival service to let oversea tourists feel this country's tourism service can provide the most excellent service arrangement to compare other countries' travelling services. When, service robotic can be participated to the country's overall tourism related services, e.g. service robots will be faster than human employees, robots will deal with calculations better than human employees, robotic will provide more accurate information than human employees, robotic will be friendlier than human employees, robots will be more polite than human employees, robots will be able to understand guest's level of satisfaction, robot can understand more clear to any questions or orders to compare human employees.

Robotic can do special requests, they work not only in a programmed frame. Moreover, when any hotel livers can be influenced to have positive attitude towards the potential use of robotic in hotels, such as being served by robotic will be a memorable experience, being served by robots will be a pleasurable experience, being served by robots will be an existing experience, preference towards the appearance of the robots, preferences towards . To compare the human employees-robots performance ratio in a hotel, if they feel robotic performance can let them to feel better service to compare human employees, then the tourists will choose the hotel to live again, when they come back to the country to travel. So, it seems that future hotels ought may choose service robots to help them to serve their clients, if they can train robotics to let hotel guests to feel more satisfactory service feeling. Their hotel service tasks may include that the reception, multi-lingual robots is a talking dinosaur , responsible for greeting, checking in and assisting guests. At the cloakroom, a robotic aim stores luggage , and porter robots carry them to the rooms, replacing a friendly , human receptionist with a robotic dinosaur may appear questionable for hospitality aficionados, but the concept may appear to be successful.

In travel bots application aspect, it may include these service robots, customer-service bots is usually incorporated for example, in the provider's website and their functions are limited to answering basic questions and assisting the user with navigating through the home page. Facebook chatbots is more interactive than customer-service bots, allowing a possibility to enter search and booking –related data using another interface, e.g. Expedia's or skyscanner's facebook, messager bots as well as the travel AIbots is such applications still rely on instance messaging to interface with the customer , but also utilize algorithms and access to information to make recommendations, e.g. a virtual travel agent uses calendar and email information to produce personalized recommendations. In conclusion, all of the above service robots will be entertained for purposes in tourism mainly due to their efficiency and reliability asnd providing excellent service performance to compare human service employees in possible , when a number of the sbove mentional technologies are already a reality in tourism, they ought not appear counter-intuitive and unsuitable for service-encounters.

I believe automation in tourism service can bring new enjoyment exciting feeling to future dream-inspiring hospitality of the tourism -sector. We also need to believe tourists will be willing to adopt interact with service robots during their holidays and what is their perceived optimal trade off between efficiency and humanity.

# TWO

# Robots How Influence Organizational Behavioral Change

Consequently creative and social intelligence will be in even greater demand as (AI) makes in management and the workforce. This development will represent a long term trend in labor markets , one characterized by intensifying demand and reward for social skills with a growing desire for creative capabilities, managers will seek to fashion of ideas and hypotheses from inside and outside of the enterprise to shape solutions to their most pressing business problems. Thus, (AI) will influence overall organizational team members who have chance to participate any decision to make more accurate business judgment.

Many managers mistakenly view judgment work as only an individual discipline, failing to appreciate that it can also involve decide interpersonal and organizational practices. In more complex settings, judgment is typically a collective outcome of individuals‘ and teams’ diverse perspectives, insights and experiences. And often , the resulting choices are better informed than decisions that an individual would have arrived at on his or her own.

Thus, when any organizations apply (AI) technology to assist managers to gather data and ideas to make any judgment. In these cases, organizations can create the conditions for effective collective judgment by establishing structures , such as " shadow advisory boards" that prompt managers and employees to source and synthesize multiple perspectives. Thus, a traditional organization (firm) might freshen its thinking is t put together a shadow advisory board, comprised of young, digital people who can apply (AI) machine assistance to make judgment work more accurate whether related to people development, problem-solving or strategizing and innovating for considerable degrees of creative and social intelligence.

Thus, on the one hand, (AI) technology machine augmentation and automation can give these advantages to human (organization managers) , e.g. developing people and community, solving problems and collaborating, coordinating and controlling work, shaping strategy and leading innovation. Besides, on the other hand, the next generation managers need have these individual attitude to treat intelligent machines to be as colleagues.

When, judgment is a human skill, intelligent machines can accelerate human learning that supports it, assisting in data -driven simulations, scenarios and search and discovery activities. Focuses on judgment work, some decisions require insight beyond what data can tell them. This is the sweet sport for human judgment, the application of experience and expertise to critical business decisions and practices. Thus, managers will also need to find ways to learn how to use digital (AI) technologies to tap into the knowledge and judgment of partners, customer external stakeholders and role models in other industries after the (AI) machine had been implemented to the organization.

Future works change:Automation, employment
and productivity

3.1 How (AI) influences employment

Human future " micro to macro" industry trends will be affected business strategy and public policy by (AI) technology. In the future (AI) technology will influence those six themes: productivity and growth, natural resources, labor markets, the evolution of global financial markets, the economic impact of technology and innovation and urbanization. However, (AI) technology will bring economic benefits of tackling gender inequality, a new global competition, Chinese innovation and digital globalization.

Nowadays, advances in robotics artificial intelligence, and machine learning are in a new age of automation, as machines match or outperform human performance in a development to any countries. For example, automation of activities can enable businesses to improve performance by reducing errors and improving quality and speed, and in some cases achieving outcomes that go beyond human capabilities. For example, some research indicated automation could raise productivity growth globally by 0.8 to 1.4 % annually; more than 2,000 work activities across 800 occupations. When less than 5% of all occupations can be automated using demonstrated technologies about 60% of all occupations have at least 30% of constituent activities that could be automated. Many occupations will change that will be automated away: Activities most susceptible to automation involve physical activities, in highly structured and predictable environments, as well as the collection and processing of data. They are most prevalent in manufacturing , accommodation and food service and retail trade and include some middle-skill jobs. For example, such as natural language processing is a key factor. Beyond technical feasibility, the cost of technology competition with labor including skills and supply and demand dynamics, performance benefits including and beyond labor cost savings, and social and regulatory acceptance will be affected by (AI) automation technology. Thus, (AI) automation will impact to influence global employment in those aspects as below:

Firstly, assuming that people are displaced by automation will find other employment. The anticipated shift in the activities in the labor force is of a similar order as the long-term shift away from agriculture and decreases in manufacturing share of employment. Both of manufacturing and agriculture industries which would be accompanied by the creation of new types of work not foreseen at the time.

Secondly, for business, the performance benefits of automation are relatively clear. Thus, the businessmen have opportunities for their micro economies to benefits from the productivity growth potential and macro economies to benefit to encourage continued progress and innovation , investment and market incentives. At the same time, employers must innovate policies to help workers and institutions adapt to the impact on employment.

This will likely include rethinking education and training, income support and safety nets , as well as support for those dislocated, when employees need to leave themselves homes to move to other cities to learn new (AI) automation works. Thus, individuals in the workplace will need to engage move comprehensively with machines as part of their everyday activities, and acquire new skills that will be in demand in the new automation age.

Consequently , the scale of shifts in the labor force over many decades that automation technologies can be a similar order to the long -term technology -enables shifts in the developed countries' workforces away from agriculture in the 21 th century. Those shifts did not result in long-term mass unemployment because they were accompanied by the creation of new types of work not foreseen at the time. However, human will still be needed in the workforce when the total productivity gains are caused by (AI) technology.

3.2 What occupations will be influenced by (AI) technology.

In the future, scientists predict that these occupations will be influenced by (AI) technology mostly. They include : retail salespeople, food and beverage service workers, language or translation teachers, health practitioners. Since these work activities have a more relevant occupations are made up of a range of activities with different potential for (AI) automation . For example, a retail salesperson will spend more time interacting with customers, stocking shelves , or ringing up sales. Each of these activities is distinct and requires different capabilities to perform successfully.

Thus, these job activities have similar simple control characteristics. Simple activities include greet customers, answer questions about products and services, clean and maintain work areas, demonstrate product feature process sales and transactions. All these activities can have similar simple activities in order to (AI) machines can be learn how to do these activities from (AI) technology . For example, the capability perception includes sensory perception, cognitive capabilities, such as retrieving automation, recognizing known patterns( supervised learning), logical reasoning problem solving.

Thus, (AI) machine is such human, which has feeling and emotion, such as social and emotional sensing, judgement reasoning methods, natural language understanding and physical capabilities, such as mobility , navigation, gross motor skill, fine motor skills. It seems that the future, (AI) human invents machines which will have these human characteristics to do human similar behavioral job duties more easily and efficiently. It implies these above human occupations will be replaced by (AI) human invention machines in the future. Due to (AI) creation, it is possible to cause unemployment number of these above workers will increase because (AI) machines can do their similar job behavioral activities.

Consequently, employers won't need to employ many of these skillful labor.

Otherwise, they can buy less number (AI) machines to attempt to do whose job activities more easily and efficiently. So, it seems (AI) machines will have more high work performance to replace these occupation workers' work performance. Finally, these occupation worker unemployment number will only increase when the (AI) machines had been invented to achieve to do their work behavioral activities absolutely success in the future.

### 3.3 Whether (A) technology machine labor will replace human worker more or assist human worker more

There is no single agreed definition of a robot how outcome of a task that is completed without human intervention. When some definitions require the task to be completed by a physical machine moves and respond to its environment, other definitions use the term robot in connection with tasks completed by software , without physical embodiment.
However, to answer the question : Whether (AI) technology machine labor will replace human worker more or assist human worker more. I shall indicate some examples to let readers to judge whether (AI) technology can create new jobs or reduce old jobs.
Firstly, I shall explain what (AI) function is. (AI) is a service robot that performs useful tasks for humans or equipment excluding industrial automation application . Thus, the classification of a robot into industrial robot or service robot is done according to its intended application. It is also a personal service robot or a service robot for personal used for a non commercial task, usually by lay persons . Examples are domestic servant robot, and pet exercising robot. It is also a professional service robot or a service robot for professional used for a commercial task, usually operated by a properly trained operator. Examples, are cleaning robot for public places, delivery robot in offices or hospitals, fire-fighting robot, rehabilitation robot and surgery robot in hospitals. Thus, these functions will be future (AI) application to our daily life necessaries or business necessaries.
However, some authors agree (AI) will bring negative outcomes of automation, due to raise competiveness, reduce human job nature. Otherwise, other authors argue (AI) will bring positive outcomes of automation, due to raise productivities, job creation, assist humans work.
On the positive outcome hand, robots can increase productivity . This is

particularly important for small-to medium sized businesses both are in developed and developing countries economies. It also enables large companies to increase their competitiveness through faster product development and delivery. Increased use of robot is also enabling companies in high cost countries to re shore, or bring back to their domestic base parts of the supply chain that will have previously outsourced to sources of cheaper labor. Currently , the greater threat to employment is not a automation, but an inability to remain competitive. Automation has led overall to an increase in labor demand and positive impact on wages. The reason is that the middle-income/middle-skilled jobs have reduced as a proportion of overall contribution to employment and earnings leading to fears of increasing income inequality, the skills range within the middle income bracket is large. Thus, robots are driving an increase in demand for workers at the higher -skilled and with a positive impact on wages. This issue is how to enable middle-income earners in the lower-income range to unskilled or retain. Finally, the (AI) positive impact supporter who argue the future will be robots and humans can work together.

However, on the negative outcome hand, robots can substitute labor activities, but don't replace jobs. They believe that less than 10% of jobs are fully automatable. Increasingly , robots are used to complement and augment labor activities, the net impact on jobs and the quality of work is positive. Automation can provide the opportunity for humans to focus on higher-skilled, higher-quality and higher-paid tasks. Robots can improve productivity when they are applied to tasks that which perform more efficiently and to a higher and more consistent level of quality than humans. For example, increased productivity is enabling some firms, such as Whirlpool, Caterpillar and Ford Motors company in the US restructure their supply chains, bringing back parts of the manufacturing process to the country of origin. Thus, productivity gains due to robotics and automation are important not just at the company level, but also for build industry and nation competitiveness.

I suppose that productivity can be raised. What are the impacts of robots on employment? Firstly, the main focus of development has been on personal entertainment, which does not drive worker productivity ( manufacturing production). When the internet ( information and communication technology (ICT)) innovation. This is borne and by findings that manufacturing productivity, which has been driven by innovations in automation rather than consumer technologies, has government strongly than productivity in the services sectors of the economy in most nature economies. It seems (AI) automation will create many jobs in internet communication entertainment game industry. For example, many young people like to use internet to play any electronic games from computer or mobile at home or outside home conveniently. Thus, (AI) automation will increase demand to be invented to any new entertainment game from internet channel. It will need to employ many (AI) entertainment game inventors to create many automation entertainment games. Thus, (AI) automation in internet entertainment game industry will need human (AI) entertainment game inventors to invent the knowledge-based capital of (AI) automation entertainment games. The (AI) entertainment game inventors will need own research and development skills, form specific skills, organizational know-how skills, databased knowledge, design and various forms of intellectual property to do these (AI) automation entertainment game invention occupations in the future.

International Federation Of Robotics(2016) indicated that China will be as a major robotics manufacturer and user of robots, benefiting from jobs created by robot manufacturing and productivity gains from robot use. Chins had sold of robots to any one single market every year since 2017 year. The Chinese government has included a focus on robotics in its 10 year strategy. In order to achieve its target of a robot density of 150 units per 10, 000 workers by 2020 year. Thus, Chinese companies will have to install around 650,000 new industrial robots between 2016 to 2020 year, 2.5 times more than installed globally in 2015 year.

Hence, China (AI) manufacturing industry will need to employ many workers . It implies (AI) manufacturing industry will create many new occupations in China. Also, ministry of economy, trade and industry (2015) also showed that Japan currently has the largest stock of industrial robots in operations, primarily in the automation industry. Driven by a rapidly aging population and low productivity rates, the Japanese government has sights on a 20-fold increase in the use of robots in the non-manufacturing sector and a three-fold growth rate of labor productivity in the service sector both by 2020 year. Thus, it also implies Japan will need many robots to be provide to service industry. Due to robots will provide to serve any businessmen's clients. Thus, it is possible that the service workers won't be dismissed as well as it is depended on the serving job nature to decide whether Japan's service workers can still serve to their employer when the service (AI) robots are applied to whose employers.

Consequently, it seems that (AI) can create employment, Ministry of economy, trade and industry (2015) showed that such as China will develop the major (AI) automation manufacturing industry. The (AI) employers will need to employ many workers to manufacture any these different kinds of (AI) robots to satisfy China or overseas individual or business buyers needs. But, (AI) can also cause unemployment to the low skillful service workers. Such as if Japan some service businesses choose to buy any (AI) service robots to replace their service staffs to serve their clients. It is possible that the service staffs will be dismissed, due to (AI) robots can do such as their same service job duties to achieve better service performance.

Thus, today, it is increasingly common for people to use robots in various situations at home and in retail stores, hotels and hospitals these service industries. Robots are classified into server types based on their functionality ( service and utility robots or those designed to communicate with humans) and appearance ( humanoid robots or mechanical robots). The type of robot, to which each country allocated particular importance in the advance of robotics, reflects the sense of values and preferences of its population. Thus, if the country has high population needs to use robots, then they will influence either more new jobs creation or more old job loss in the country's (AI) manufacturing or (AI) service industries both. For example, Japan respondents often associate the term " robot " with humanoid robots that can communicate with human and they have a high level of familiarity with robot. The US has the highest level of robot utilization at home and in retail stores with its people being the most enthusiastic about the future use of robots. Germany shows a strong tendency to consider robots for industrial purposes and its people feel strong effort to the presence of robots in their households.

In conclusion, to judge whether how (AI) will influence the country's employment to be better or worse. It will depend on the country home buyers (users) or business buyers (users) how to use (AI) for their daily needs. If the country , such as US retail stores need to use (AI) , it will have possible to reduce some or many retail service workers. Even, if the country , such as Japan has many home users need to use (AI) , it will not influence the employment market. Otherwise, it will raise (AI) salespeople numbers. Even, if the country, such as Germany and China will have many (AI) manufacturers, then it will create many (AI) manufacturing occupations for these (AI) manufactory workers.

Consequently, (AI) robots manufacturing and service needs will have positive or negative impact to any country's employment. It will depend on the (AI) service provision and service workers' job nature as well as the manufacturing workers of (AI) knowledge level to decide their employment chance in their country's employment market.

● Could work activities in China be automated
making in the nation with the world's largest automation potential?
Can (AI) technology influence China economy? Could China workers be affected and jobs made up of routine work activities and predictable? Will programmable tasks be particularly impact to China employment market ? When impact on labor market is likely to be gradual at the aggregate level, it can be sudden and dramatic at the level of specific work activities, rending some job obsolete fairly. Overall (AI) technology will raise digital skills when reducing demand for medium incomer inequality for China workers. It seems (AI) technology's effect on productivity could be crucial to China's future economic growth as the population ages are increasing.

In China, some biggest technological companies driving significant investments in research and development. Moreover, China is one of the leading global (AI) technology development county. However, China will need to focus on building its innovation capacity. For example, United States and United Kingdom are currently producing more influential (AI) technological research. However, if China planed to achieve (AI) technology success, it's traditional industries will need to develop technical know-how –to and overcoming implementation costs prepare to develop (AI) . When (AI) technology is introduced into China society, China government needs to raise concerning ethical, legal, technological security etc. business questions. Also, surrounding issues include privacy, discrimination, legal liability and regulation. It aims to encourage overseas investors to choose to invest (AI) technological industry to raise GDP growth and manufacturing industries income growth for long term in China.

If China encouraged overseas (AI) technology investment in its country. It is possible to influence China employment market to be changed. Because (AI) technology will impact to influence China people daily life. Due to (AI) technology is introduced to China society, many rich people will prefer to spend to buy any high (AI) technological products for entertainment or learning or machine man driving etc. daily necessity activities. Then it will raise GDP growth and will raise (AI) manufacturers or related-(AI) technological manufacturers profit. It is beneficial to China because it can become one high knowledgeable and (AI) technological economical society.

But it will bring bad influences to raise unemployment chance for the low skillful labor. In labor economy aspect influence , how (AI) technology can influence China low skillful labor unemployment ratio raising. The raising low skill labor unemployment reason is because China low skillful human labors are argued or are replaced by (AI) technology creating new challenges to introduce to influence China society of simply human manufacturing job nature to be changed to be high (AI) technology manufacturing job nature in any China factories. Moreover, when (AI) technology introduction to China, it will cause other related social challenges in China. The varied (AI) related challenges, including the difficulty of creating safe and reliable hardware for sensing and affecting ( transportation and education), the challenges of gaining public trust, a low resource comities and public safety and security, the challenges of overcoming fears or marginalizing humans in China employment and workplace and the risk of diminishing interpersonal trust because the low skillful labors won't believe any China employers will give chance to employ them , due to (AI) technology will replace their skills and man manufacturing of productivity is much less to compare to (AI) technology manufacturing method.

- How does ( AI) technology influence
the future of employment change?

Are future nature of jobs changed to computerization from (AI) technology? Where are the probability of computing occupations from (AI) technology influence? What is expected impacts of future computing on labor market from (AI) technology influence? John Maynard Keynes's frequently cited prediction of widespread technological unemployment " du to our discovery of means of economic the use of labor outrunning the pace of which we can find new used of labor" ( Keynes, 1933, p.3).

In the future, (AI) technology will impact some nature of occupations to change computing. This chance will also influence some countries' economic change. For example, some factory human labors hand routine manufacturing tasks will be changed to computerization of routine manufacturing tasks by (AI) technological machine men hand manufacturing method. it will cause a structured shift in the labor market, with workers reallocating their labor supply from middle-income manufacturing to low-income service occupations.

Arguably, this is because the manual tasks of service occupations are less computerization, as who require a higher degree of flexibility and physical adaptability. So, (AI) technology will influence the human hand labor skillful occupation nature of task cheaper , such as vehicle manufacturing , ship manufacturing, computer manufacturing, steel manufacturing, television, radio etc. home electronic products of heavy machine industry change. Due to (AI) technology machine man will be proper to be used to manufacturing these electronic products when the (AI) technology innovation can develop to the mature stage. Then, any countries manufacturers will choose to use (AI) technology machine man, instead of human hand production.

Supposing the future prices of computing are fallen, seriously, problem solving skills are becoming relatively productive, explaining the substantial employment growth in manufacturing occupations, involving cognitive tasks where skilled labor has a comparative advantage, as well as the increase education needs for (AI) technology computing of machine man subject study.

Prediction of education needs for (AI) technology student numbers will increase, due to manufacturing industry needs many (AI) technology students in future employment market. Another (AI) technology influence if the future (AI) technological innovation, e.g. machine man manufacturing or machine man service industries will both increase demand, then with more sophistic software technologies will be disrupted labor markets by marketing workers redundant.

For publishing industry, what is striking about the case in paper book publishing industry will be unpopular? Due to the electronic book publishing industry will be popular, e.g. Amazon publish . (AI) technology can influence paper book manufacturing method which is replaced by machine man electronic book manufacturing method as well as it will cause the computerization is no longer confined to routine manufacturing tasks. Due to (AI) machine man manufacturing technology will be proper to be used to manufacture any products in short time efficiently and effectively , e.g. electronic book products. In the future, if it is fact to occur this case, such as ( AI) technological machine man manufacturing method will be adopted ( applied) to manufacture electronic books or any products in possible. (AI) technology will cause many manufacturing workers are unemployed. It is beneficial to employers, who can reduce to spend much wages expenditure to employ manufacturing workers, but it will cause many manufacturing workers loss jobs and reduce income to support whose families lives. It will cause social challenges, e.g. increasing stealing crimes if the manufacturing workers had not other skills to find other jobs to do easily. So, manufacturers need to concern over technological unemployment which will be hardly future phenomenon if who decided to dismiss all manufacturing workers, due to (AI) technology machine men replace to them.

If ( AI) technology can be innovated to produce any kinds of machine man to serve any service or manufacturing industries successfully. Then, it will bring these questions: Can future that workers be influenced to be automation employment and productivity by (AI) technology influence? Does it impact to influence the (AI) technology countries' productivity and growth and natural resources development and labor markets and evolution of global financial markets and economic impact of technology and innovation and urbanization etc. issues? How will automation transform the workplace? What will be the implication for employment? What is likely to be its impact both on productivity in the global economy and on employment?

In fact, automatic of activities can enable businesses to improve performance by reducing errors chance and improving quality and speed, and same cases achieving outcomes that go beyond human capabilities. Some economists indicate (AI) technology would give a needed boost to economic growth and prosperity have of the working age population in many countries. Based on the scenario modeling, they estimate automation could raise productivity growth globally by 0.8 to 1.4 % annually. They also indicated that almost half the activities people are almost $1.6 trillion in wages to do in the global economy have the potential to be automated adapting current demonstrates technology, according to their analysis of more than 2,000 work activities across 800 occupations. When less than 5% of all occupations can be automated entirely using demonstrated technology, about 60% of all occupations have at least 30% of worker made activities, that would be automated. More occupation will change to be automated. They also indicated for business performance benefits of automation are relatively clear, but the issues are more complicated by policy making to attract foreign investors. Beyond technical feasibility, the cost of technology, competition labor will include skills and supply and demand dynamics, performance benefits and beyond labor cost savings and social and regulatory acceptance will affect the automation. Their predictions suggest that half of today work activities could be automated by 2055 year, but this could happen 10 to 20 years earlier or latter depending on the various factors in addition to their wider economic condition.

Some scientists suggest (AI) technology is finally starting to deliver real-life business benefits. Computer power is growing significantly , algorithms are becoming more sophisticated and perhaps most important of all, the world is generating vast quantities of the fuel that powers (AI) technology data billions of gigabytes of it every day. Also, online firms are digital natives, such as Google online search service company is investing on (AI) technology. For new though most of the news if coming from the suppliers of (AI) technologies. And many new users are only in the experimental phase. Few products are on the market or are likely to arrive these soon to drive immediate and widespread adoption. As a result, analysts believe (AI) technology's potential will give true economic benefit in the future. (AI) industry will introduce to suppliers and users to raise economic potential of (AI) technology.

In the future, (AI) technology systems can solve business problems. Some scientists categorized those into five technology systems that are key areas of (AI) technology development: robotics and autonomous vehicles, computer vision language virtual agents and machine learning , which is based on algorithms that learn from data without replying on rules-based programming in order to draw conclusions or direct an action.

Such as computer vision and language includes natural language processing, analytics, speech recognition technology, some are about learning from information, such as about machine learning and others are related to acting on information, such as robotics, autonomous vehicles and virtual agents, which are computer programs that can converse with humans. Machine learning and a subfield called deep learning are artificial intelligence applications.

- Can artificial intelligence impact
global economy growth?

Artificial intelligence (AI) is a term first defined in 1956 year. It is a branch of computer science that aims to create intelligent machines that work and react like humans. In contrast today, 60 years later, (AI) is characterized by a number of applications, including computers playing games against humans and understanding human languages, virtual personal assistants, and robotics which involve computers seeing , hearing and reacting to sensory stimuli. In the future, technologists predict for (AI) technology ranging from (AI) being used as a tool to aid relatively simple processes for robots with human like mental capabilities, who expect (AI) technology can emulate human performance by learning, coming to mind its own conclusions, understanding complex content, engaging in dialog with people, enhancing human cognitive performance or replacing humans in executing both routine and non-routine tasks. In existing industry, (AI) technology is used , such as targeted advertising and virtual used personal assistant as well as the (AI) technology that my exist in the future, such as robots with human vehicle processing capabilities.

The range of (AI) technology's progress in the future will determine the economic impact future of (AI) technology on the global economy with more limited advances and applications ( i.e. weak (AI) only) corresponding to more limited economic impacts and more substantial progress, i.e. strong (AI) technology is corresponding to more significant economic impact.

(AI) technology learning that automates analytical model, including predicting cause-and-effect relationship from biological data, identifying new drugs, self-driving cars and protecting against fraud etc. functions. Also (AI) learning can improve natural language processing that allows computers to continue to better analyze, understand and generate language to interface with human using the natural human language, virtual personal assistant, helps users by providing scheduling appointment, reminds organizing personal finance and finding providers of various services, machine vision allows (AI) machine man to identify object, scenes and activities in detect pedestrians and bicyclists.

We expect the economic effects of (AI) technology to include both direct GDP growth from sectors that develop or manufacture (AI) technology and indirect GDP growth through increased productivity in existing sectors that employ some from of (AI) technology. If (AI) producing sectors could grow, then it could lead to increase revenues and employment of (AI) technological professionals within these existing firms as well as the potential creation of entirely new economic activities to any countries' societies productivity improvement in existing sectors could be realized through faster and move efficient processes and decision making as well as increased (AI) technological knowledge and access to information available in societies easily.

In the future, if (AI) technology is an increasingly critical component of more products, it will become an integral part of necessary products of many people's lives. The extent of (AI)'s economy effort is also likely to vary from region to region, thought variation may be more dependent on the predominate economic activity of a region and the (AI) ability can influence economic activity, rather then the economic or developmental status of the regions. (AI) technology can move accessibility and can use source development to do international business between one country and another country.

So (AI) technology has the potential to give benefits to different income chooses and to bring significant gains to both developed and developing countries. For agricultural technology, (AI) has the potential to optimize food production around the world by analyzing agricultural regions and identifying what is necessary to improve crop yield. In total, (AI) technology gives greater economic impact to any countries agricultural regions if which implemented (AI) technology to grow crop , fruit etc. food production in the farms.

Investment in (AI) technology is such as capital investment to any countries' public or private enterprises. So, it will have large economic impact to the future . If the (AI) technology is reasonable invested to the different needs aspect by the public or private enterprises in the country. Then, it will have good economic impact to the country in the future. However, when (AI) technology is likely to affect both the productivity and employment components of economic growth in many sectors. Significant public debate has focused on projections of (AI)'s effect on the labor force. However, for instance, some researchers have argued that the rise of (AI) technology and automation will led to significant unemployment as capital is substituted for the low skillful labor. So, they point to the concern that the increasing sophistication of (AI) technology may balance skilled and semi-skilled workers and the reduce the size of the middle class. However, this is not a new argument, due to (AI) technology negatively affecting the labor force and leading to mass unemployment. Because the (AI) technology is the substitution of machinery for human labor. Although, employment in certain industries, has been reduced in the past due to technological advancement. For long term, the labor market has adapted to the introduction of new technology, giving rise to new jobs in new areas. (AI) technology may also be accomplished without a reduction to total employment in the long-term to some Asia countries, such as Hong Kong and Japan. Because Hong Kong and Japan many low skilled labor, e.g. security, cleaner who complaint that employers need them to work long time hours. ( abnormal working hours) e.g. one day 12 to 15 working hour per day. Hence, if (AI) machine means invention technology success. Security or cleaning job can be worked from (AI) machine man in some hours every day in order to reduce the long time working hours cleaners or security workers, e.g. one (AI) machine man works 4 hours for cleaning or security job, one day as well as another cleaner or security labor only needs to work 8 hours one day. So total security or cleaning employers can employ 12 hours machine cleaners or security workers and human cleaners or security workers in one day. For long term benefit, Hong Kong or Japan every security or cleaning worker does not need to work 12 hours minimum working hours one day. They won't feel tried and bore and without private with whose families, so who will accept to do these cleaning or security jobs, even they can raise work efficient and performance when who feel happy and health.

So, (AI) technology of machine man invention can raise low skillful labor efficiency and it can help them to avoid abnormal working hours demand in some busy work life countries, such as Hong Kong and Japan. Before, one Japan female labor feel unhappy to work, due to who often needs to work abnormal working hours for her employer and who has less sleeping and without any private time to enjoy her life with her families every day. So this abnormal working hours factor causes her to do commit suicide behavior, then she is die unlucky. So (AI) technology of machine man invention ought avoid abnormal working hours demand for employer in any countries in the future.

The most important occurrence to any employers, some researchers had attempted to do one experiment to find that private research and development , venture capital and public research and development investment all have strong net effect or economic growth with venture capital funding further having the strongest such effect from (AI) technology. The researchers hypothesize the venture capital investment contributes to economic growth through (AI) technology innovation and by the capacity of an economy to use existing (AI) technology knowledge to increase productivity. They predict the impacts of venture capital, business-research and development and public research and development can raise multi factor productivity from (AI) technology introduction.

Can (AI) technology influence the economic development to developing countries? The developing regions of the world contain most of natural resources. If one day, (AI) technology has invent one kind of machine man which can assist any gas or oil workers to seek any new oil/gas natural resource locations easily. I believe that (AI) technology can help these natural resource exploitation countries will gain economic benefit more easily. So, (AI) driven technology can be used to change to create any new opportunities to address poor management or resources and improve human well being, such as Africa Latin America and India can use (AI) technology machine man to seek any oil/gas natural resource countries exploitation activities to attempt to gain much economic benefits.

- Why will (AI) technology grow economic

development ?

Nowadays, increases in capital and labor are no longer driving the levels of economic growth, such as (AI) technology. The ability of increase in capital investment and in labor of traditional drivers of production, have no longer to be enjoyed in most developed economies ,e.g. developed country, US, UK . However, artificial intelligence has the potential to overcome the physical limitation of capital and labor to avoid missing out on this opportunity. So, policy makers and business leaders must prepare for and work toward a future with artificial intelligence. They must do with the idea that (AI) is another simply method to enhance productivity method . Rather they must see (AI) as the tool that can transform thinking about how growth is created.

Economists have always thought of new technologies are as driving growth their ability to enhancing. It can replace labor and capital factor of production. So, it brings this question: What is the factor of production (AI) technology characteristics. They key factor is to see (AI) technology as a capital-labor .

(AI) can replicate labor activities at much greater scale and speed, and to even perform some tasks began the capabilities of human. For example, by using virtual assistants , 1000 legal documents can be reviewed in a matter of days instead of taking three people six moths to complete. Some (AI) technology may be one kind of factor of production in the future. For another example, people will work in workplace digitalization environment. So, in the future, working environment and information management are automated. Such as Konica camera sale company will use workplace digitalization. So , (AI) technology can provide workplace digitalization in order to raise productivity efficiency. (AI) technology will be one kind of production which is replaced by workplace digitalization and it will grow any organization productivity efficiently. Then, (AI) technology will assist overall social economy growth , due to productivity is raised and products can be produced in short time to prepare to sell in consumption market. So, time will be shortened to increase GDP growth fast for the development of (AI) technology countries.

- How can (AI) technology impact to global

economic and social and psychological
changes?

What will be the development of (AI) technology and predictions concerning the future evolution? The computers and robots will develop conscious, intelligent and minds into humans, enhancing psychological and behavioral abilities and allowing for direct communication with (AI) minds. (AI) technology will be impacted human life by (AI) technology information communicative and environmental influence. A " world brain" and " world mind", this psychological system will be enhanced and enriched the capacities of both individual and collective cognition by (AI) technology of service industries.

(AI) technology with influence these human needs of service industries changes, such as , biological science, finance, entertainment, business, biological science, transportation, communication military etc. The personal computer evolution, the internet and the world wide web which exploded on the scene, linking business, homes, schools, social organizations which were a completely unpredicted phenomenon to influence human life. Kurzweil (1999) predicts that by 2029 year, most human communication will be with machines. According to Person, by 2100 year, there will be human machine convergence.

How can (AI) technology influence environmental protection to make benefits to farming economic growth? (AI) technology can be applied to predict how to solve environmental pollution challenge to avoid to damage any crop or vegetable or rice or fruit etc. food growth. Because environmental experts can gather global environmental pollution data from an environmental database to build a perform a systematic analysis from (AI) technology. The first step is this broad analysis can include understanding, statistical and data gathering techniques to obtain the relevant data, the correlation among the variables involved, and a list of possible models. The next step is to select a set of methods and models that cover all kinds of knowledge and functionalities needed for the decision making process. Once the models are selected, they must be fully implemented by means of machine learning , data mining, statistical or numerical technique. After that, those models must be integrated to build the whole EDSS. The EDSS must be tested to check its performance, accuracy, usefulness and reliability, both from the user's and (AI) technology/computer scientist's point of view. If these is any wrong feature in any development stage, such as model's integration, models' implementation, selection of models, database, problem analysis etc. the developers must come back in the update th required components. When the evaluation phase is all right, the EDSS is ready to be applied to the environment. The great contribution of artificial intelligence to EDSS the integration of several methods complementing the classical statistical models/simulation , statistical analysis, linear models, etc. and numerical models ( control algorithms, optimization techniques etc.) .

This cooperation makes the resulting systems more reliable and powerful in coping with real world environment systems. Date interpretation has been a principal area of research in (AI) technology since the very beginning. The most demanding problem in the environmental assessment context. Knowledge representation permits the definition of the different types of data that the existing methods adapt to the process. There is also a lot of work to clean, repair and transform the huge available quantities of raw data. Apart from this, the availability of meta-information or background knowledge is required to guide the process. Data mining is multi-disciplinary: It covers expert systems, data based technology, statistics, data visualization and unsupervised machine learning. These techniques operate at the level of data and background information, where numerous and often incompatible new commensurate pieces of information from disparate sources have to be brought together ( K, Fedra, 1994).

So, it seems that in the future, (AI) technology with the increasing maturity in particular those related to knowledge and engineering, new dimensions can be assisted to users in environmental decision making are available. For example, many environmental systems are characterized both by incomplete models and by limited data. Hence, in the future, (AI) technology will be applied to predict climate change to reduce crop or fruit etc. food agriculture challenge by climate change bad influence.

- Will (AI) technology influence digital economy change to manufacturing industry ?

To understand how the manufacturing business must adapt to prosper in the technology, we need to understand how (AI) technology will change us to shape our daily habits to satisfy our expectation of products to how we shop and even the immediate of the entire process. For example, taxi services are in the crosshairs as on demand transportation services like, available of the touch of a smart phone button expand. In fact, Yellow lab, US country , san Francisco city's largest taxi company is filing for bankruptcy as the industry starts to change faster than almost anyone expected. However, at this point, its more than an app that is changing, some our taxi passengers renting taxi transportation to catch consumption behavior.

(AI) technology will influence digital economy for taxi passenger's individual customer experience, offering a growing renting taxi to catch of service and feedback opportunities when any one taxi passenger who chooses to use mobile phone app online tool to prepaid to rent any taxi more easily.

Also in the long term, (AI) technology can influence vehicles drive themselves of behavior. Already, companies like Google and GM are working on projects to bring fleets of autonomous vehicles to cities at the path of a button.

Moreover, this on-demand service model is beginning to appear across a much broader range of markets. For example , Amazon company is investing in its own fleet of trucks, planes and even drone at the same time as it pushes for same-day delivery of products. As some point, vehicles will be autonomous too. So, it seems that (AI) technique will influence any transportations choose to use digital autonomous driving technology in the future . For Amazon company case, it is not stopping of logistics. It is also aiming to automatically manage the supply of consumer home products with its recently launched Amazon replenishment service, Dash. Dash is a digital service that enables that connected derive to automatically order physical products from Amazon when supplies are running low. So, it seems (AI) technology will be applied to logistic function by digital technology method introduction in the future.

Hence autonomous vehicles will optimize industry supply chains and logistics operations through increased efficiency and flexibility. In fact, fully automated and lean supply chains will keep reduce load sizes and inventory by leveraging smart distribution technologies and smaller autonomous vehicles by machine man assistance. If Amazon continues to grow market share for online sales by reducing effort required by the consumer to place an order, when also contributing the almost immediate delivery of products to the doorstep. So, it will further fuel the trend toward on-demand derive. As Amazon company fuels the on-demand economy, consumers will expect immediacy in more parts of the digital economy. On top of speed, consumers increasing expect more personalization options.

So, (AI) technology will influence digital manufacturing, such as Amazon publishing to monitor every aspect of every process in real -time and communicating to self-optimized deep learning robotics, new methods of high volume and high customization will become possible. Then, as products merge into product platforms and even services, manufacturers have the opportunity to provide components and platforms used by smaller players. So, (AI) technology will influence manufacturing industry to choose automated SMI lines, robots installed, automation engineers.

Another future (AI) technology development can be applied to space science aspect, such as Automation engineering space in manufacturing process to achieve digital manufacturing benefits to any businesses in the future. Such as reducing cost, shortening manufacturing time, raising efficiency, shortening delivery products to client individual time. How can artificial intelligence give the need and advanced fast and evaluation methods benefits for space exploration? When US NASA ( space exploration organization) achieves any space exploration missions, it will answer this question:

When is it useful to have a machine use (AI) technology to achieve a decision? After all, after millions of years of space exploration and rough 10,000 years of civilization, humans are usually quite good at making decisions in complex uncertain environments. Through, Johns Hoplains University's Applied Physical Lab. Research in (AI) technology enabled systems, which has identified three general use cases for (AI) technology to explore space mission:

First, for some tasks (AI) technology is more cost effectiveness than human. Second, (AI) technology is better suited than humans at solving some, but not all problems. Third, (AI) technology allows NASA organization's space exploration mission to develop machines that ate capable of responding faster than when a human is in the decision loop ( D. Scheidt, 2012, A. Castano et. al. 2008).

So, the use of (AI) technology to enable science by observing the pace of rapidly evolving phenomena was demonstrated. It is more effectively coordinating and (AI) technology utilizing to earn economic benefits to use for space exploration mission.

However, (AI) technology also have current risk for space exploration. Today (AI) technology is immature and requires further development to reach its potential. For instance, the (AI) technology algorithms that detected the dust derive could not have identified whether the Martain weather represented a threat to the cover. Also it can not yet use instrument input to determine what, where and how to autonomously make the next space science measurement. An equally important factor limiting (AI)'s deployment is that lacks the methodology and technology to effectively test (AI) technology. So, the challenge will testing (AI) enabled system is how (AI) performance can be measured. It would be NASA organization's difficulty to find (AI) technology to develop to carry on researching any space exploration missions in the future. However, (AI) technology will be a good economic benefit choice for space exploration mission in the future.

- What is artificial intelligence potential
benefits and ethical considerations?

The ability of (AI) technology systems to transform vast amounts of complex information into insight has the potential to help solve manufacturing or service challenges for human needs. However, to reap the societal benefits of (AI) systems, humans will need to trust then and make sure that which follow the same ethical principles, moral values, professional codes and social norms that we humans would follow in the same scenario, research and educational efforts as well as carefully designed regulation in order to achieve the most effort of economic benefits goals. For example, international business machines corporation (IBM) is actively engaged both competitors , in global discussions about how to make (AI) ethical and as beneficial as possible for people as social economic benefits.

(AI) is usually defined as the " capability of a computer program to perform tasks or reasoning processes " that human usually associate to intelligence in a human being. Often, it has to do with the ability to make a good decision, even when there is uncertainty, too much information to handle. As an example, play chess or complex card games of entertainment activities is believed to need some form of intelligence in a human being, as well as choosing the best medical facilities in a difficult medical case, or creating something new, such as mathematical theorem or even some form of act, or even driving automatic machine man ( self driving vehicle) replacing human driving in the middle of a crowded city.

(AI) needs depends on what we consider being intelligence in the behavior of a human being act a certain point in time. If human belief about human intelligence changes and we don't believe any longer that a certain task requires intelligence, then a computer program performing that task is no longer part of (AI), it becomes just another boring computer program. So, it means that (AI) technology will replace some old computer programs, if human can invent new generation of (AI) software for any functions or

activities to satisfy human needs.
As IBM, it argues intelligence. This means that we aim to build systems that enhance and scale human expertise and skills rather than replacing them. We therefore focus on practical applications of (AI) capabilities that assist people in performing well-defined tasks of needs by exploiting and wide range of (AI)-based services. We also use the term " cognitive computing" it is mean a comprehensive net of capabilities based on technology. It comprises the fields of machine learning, reasoning and decision technologies, language, speech and vision recognition and processing technologies, high performance and high efficient functions for any industries or individual consumers needs. For example, robotics, which are usually very good at doing what which are supposed to in any environment, much have public shopping center, factory etc. places which need simply services from the robot ( machine man), such as cleans the floor of our houses to the robot that can work together with humans in production chains, passing through the warehouse, robots can take care of the tasks of an entire warehouse and the companion robots like Nao, Pepper, Aibo and Giraff, who can entertain use, talk to use and help elderly people to stay connected to their friends, relatives and doctors.
Google company is building automatic machine ( self-driving cars ) and has acquired more than 10 robotics companies. Facebook had opened whole new research facility only on (AI) research. Apply computer has developed Siri. Microsoft computer company has built a similar personalized assistant. Google has Deep mind, a UK company whose long term aim is to build general (AI) and has already great potential to win game to the world champion and IBM is investing a huge amount of resources in applying its Watson cognitive computing system to the medical domains to finance and to personalized education. In Europe, IBM is establishing new centers in Munich and Milan focused in the application of cognitive computer capabilities to the internet of things and healthcare respectively.
For example, automatic machine man ( self-driving cars) are all about (AI), which used to be able to see what happens in the street ( signals ,lanes, other cars, pedestrians, traffic lights, which need to able predict what other cars and pedestrians will do, and who need to be able to cope with unforeseen situations. Since, most car accidents are due to human fault, it is estimated that the adoption of self-driving cars will save about half of the lives that are usually last in car accidents.
IBM Watson company has to understand spoken language, make sense of

massive amount to text , respond correctly to questions in many categories, as well as assess its own confidence in responding to such questions. In the future, (AI) technology can own question/answering capabilities that would be very useful, for example, in assisting a doctor when trying to some to the correct diagnosis for a patient and to propose the best therapy .
Intelligent machines can also rely on huge amounts of data to be used to learn how to make better decisions. This data comes from all of us over the years Facebook users have uploaded more than 250 billion pictures and every day who upload about 350 million more. Every second, we submit 40,000 google search queries. So, (AI) technology will be connected through the web from appliances to traffic lights from cars to watches. Other tasks that are very easy for humans are physical and manipulation tasks, such as walking , running, picking up an object to make its shape and location, restricted environment. But (AI) machine man technology still not able to have the general physical and manipulation capabilities even of a 6 year old.

So, it brings this question: Why do (AI) scientists need to concern ethics? Because (AI) technology is complex, information into insight has the potential to reveal long held secrets and help solve some of the world's most difficult problems. (AI) systems can potentially be used to help discover insights to treat disease, predict the whether, and manage the global economy. So, ethic issues is important to and (AI) scientists . If any one new (AI) technology research investigation could success, it will be a secret to and the (AI) scientists can not permit to their loyalty to any competitors to damage the fair (AI) technology products trading market. The country ( countries) (AI) technology scientists need to concern ethic issues, who need to keep secrets for their countries economic or/and social benefits. This is moral issues to any countries/country loyalty is whose countries intangible assets. They can not sell (AI) loyalty to any their countries to assist whose economic benefits immorally.

- How can (AI) technology influence to global
health care economy development?

According to (AI) lecturer analysis, when combined key clinical health (AI) application can potentially create $150 billion in annual savings for the US healthcare economy by 2026 year. (AI) technology is re-winning modern conception of healthcare delivery. It enables machines to sense, comprehend, act and learn. So which can perform administrative and clinical healthcare functions (Accenture, 2017).

It will help health care service organizations to reduce health care cost, will improve and raise service quality and access. So, (AI) health market size will be predicted growth. (AI) applications in health care include robot-assisted surgery, virtual nursing assistant, administrative workflow assistant, fraud detection, error reduction connected machines, clinical trial participant identifier, preliminary diagnosis, automated image diagnosis and cybersecurity.

What kind of benefits (AI) technology can contribute to healthcare service? (AI) technology can deliver what many health care organizations need, such as financial and operational of labor costs, digital expectations from patient consumers how to use (AI) technology to solve interoperability challenges in any healthcare organizations. Also (AI) technology can be applied to wellness an d lifestyle management, diagnostics, delivers financially but also way of organizational and workflow improvement. So, (AI) technology will be continue to become most prevalent and adoption to healthcare organizations , which must need to enhance structure to be position to take full advantages of new (AI) technological capabilities. (AI) technology can change the nature of work and employment is rapidly changing to make the best use of both humans and (AI) talent in healthcare industry in the future. For example, (AI) technology offers a way to fill in gaps and the rising labor shortage in healthcare. According to Accenture analysis, the physicians shortage is increasing. However, (AI) technology will manufacture healthcare machine men to replace physicians in future one day(2017). Hence, (AI) technology will be invented to raise health care service staffs work efficiency and performance in any hospitals or clinics in the future.

In conclusion, (AI) technology will raise efficiency for any service or manufacturing industries in the future, although, it is possible that it will also rise low skillful workers unemployment numbers. But, the most important influence to human technological innovation will be risen and it will influence human life will be changed to be better, e.g. self drive cars, health care physician machine men, machine man cleaners etc. intelligent machine men will be manufactured to serve for our daily life. Furthermore, (AI) technological products will influence countries trading, some low technological development countries manufacturing businessmen can choose to buy any (AI) products to raise whose productivity and efficiency and reducing cost to achieve economic cost saving result. Also, GDP of trading growth income will increase to the (AI) products sale countries.

Hence, it will be beneficial to economic development to both developed and developing countries both in the future as well as (AI) scientists time and money spending will be valued to continue to invest (AI) technology development for human life and economy benefits for long term.

In conclusion (AI) technology will raise macro economy growth and it can create many (AI) jobs , but it also raise the low level technological worker unemployment change. In the future, (AI) technology can be applied to digital technology to attempt to invent any new undiscovered (AI) and digital technology. So, it needs any scientists to continue to research how digital and (AI) technology can be mixed to satisfy human's future undiscovered needs.

Reference

A. Castano et. al. " Automatic detection of dust devils and clouds at Mars" Machine vision and applications, Oct. 2008, vol. 19, no 5-6, pp. 467-482.

Accenture, " Why artificial intelligence is the future of growth"(2017) <http://www.accenture.com/us-en/insight-a rtificial-intelligence-future-growth>.

D. Schedidt , Unmanned Air Vehicle Command And Control, Handbook Of Unmanned Air Vehicles, Springer-Verlag, 2014. Facebook (AI) Research Available at https://research.facebook.com/ai, research at google, machine intelligence available at http://research.google.com/pubs/machineintellige nce.html; micro soft research-machine learning and artificial intelligence available at http://research.microsoft.com/en-us/research- areas/machine-learning-ai.aspx.

International Federation Of Robotics, 2016. IFR press release world robotics report. IFR, org . 29 Sept. Accessed Feb. 01, 2017. http://www.ifr.org/news/ifr-press-release/world-robitics report -2016-8321.

K, Fedra , "GIS and environmental modelling" in environmental modelling with GIS, edited by M.F. Goodchild.B.O. Parks and L.T. Steyaert, Oxford University press, pp. 35-50, 1994.

Keynes, J.M. (1933). Economic possibilities for our grandchildren (1930). Essays in persuasion, pp.358-73.

Mckinsey & Company (2013, May). Disruptive technologies: Advices that will transform life, business and the global economy , USA.

Ministry of economy, trade and industry, Japan, 2015, Japan's robot strategy. Ministry of economy, trade and industry.

Ray Kurzweil , The age of spiritual machines (1999) is cited numerously through this chapter: Kurzweilai.net http://www.kurzweilai.net

Rich, Elaine & Knight, Kevin, Artificial Intelligence Second Edition, 1991, New York; Mc-Graw-Hill.

# THREE

# HOW ROBOTS CHANGE STUDENTS LEARNING BEHAVIOR

Nowadays, artificial intelligence (AI) is widely knowledge to be one kind of the dramatic technology. However, it is expected to continue, to have a disruptive impact on human's private and public life, so defense and security will be no exception. But how exactly will these be affected ? How will (AI) defense and security is incremental in nature? If (AI) technological machine men are applied to teach students in education aspect, is it better to my next generation learning develpment more than they are applied to war attack aspect.

To research why artificial intelligence (AI) has possible to be used to cause autonomous weapons by human. We need to understand these three aspects of relationship. They include cybersecurity and artificial intelligence and machine learning and autonomous weapon systems relationship between of them. Basic on (AI) machine can be invented to learn any new knowledge, so if (AI) machine men are taught how to attack enemy, which will be such as human soldier function. But, if (AI) machine men are taught how to learn university knowledge to teach students. Then, they will be such as human lecturer function. So, when (AI) is invented to own human mind and judgement and learning abilities, then they will be either human's enemy or human's assistant, such as university lecturer's assistant.

Firstly, we need to know what is the mean of artificial intelligence and cyber defense/offense? It means defense of critical networks: real time, pattern finding, anomaly seeking, it must utilize machine (AI) learning algorithms to efficiently, and instantaneously respond to potential network threats as well as it means human on or out of the loop. On the loop : it means anomaly detection: human notified, IT analysis, response. Out of the loop: it means anomaly detection: (AI) decides best method of response: quarantine, honey pot monitoring, hack-back. Thus, it is possible that (AI) can be used , such as autonomous cyber weapon. If (AI) is applied to make the decision best method of response to learning aspect, such as univeristy different subject knowledge. Then, it will be one good technological educational tool to teach university students.

In simplicity, (AI) can be one of scientific weapons platform or one of university teaching tool. When one day, it is invented to be applied to control war planes to fly to any countries to attack enemies or it is invented to be seemed to human to replace soldiers to bring guns or any weapons go to other countries to attack. So, it is possible that future any war defense planes, (AI) technological automatic control weapon can be replaced of human soldiers or war plane pilots to control any war defense planes to go to different enemy countries to attack them easily. It is very horror matter to threaten global human's ourselves life in the future , if (AI) automatic control war defense planes or (AI) automatic control machine soldiers were invented successfully. Otherwise, when one day, (AI) machine lecturer is invented to be applied to learn university different subjects knowledge to replace lecturers to copy lecturer's every prepared lecturer course to speak to let students to listen when they are sitting in university halls as well as the (AI) machine lecturer can make analysis and judgement response to answer every student's enquire immedicately after it had speaking all courses to students to listen in lecturer hall every time. Then, it can let human lecturer does any education job duty, e.g. research education work. So, (AI) machine lecturer will be future human lecturer's assistant in future one day.

Finally, the most serious (AI) technological invention risks are human is unknown these aspects of (AI) absolutely: They are not simple automatic systems, learning reasoning, communication of " self-aware" systems. Thus, human will face (AI) technological invention risks or threats if human invent (AI) machine man to learn how to attack enemy. Otherwise, if human invent (AI) machine man to learn how to teach univerity student. I believe that my future university students can raise learn ability and writing ability and reading ability from (AI) machine lecturer teaching more than human lecturer teaching.

### 1.1 Online technology and online book technology influences artificial intelligence mind development

Nowadays, online technological invention bring online book technological development. Also, artificial intelligent technological machine men had been invented to link internet to do any jobs, e.g. children can find any data from artificial intelligent machine men when the artificial intelligent machine man had been installed internet and computer function, then children can find any online books to read from the artificial intelligent machine man. Such as Japan artificial intellgent machine men had installed computer and internet function, the Japan family children can find any online books to read from the artificial intelligent machine man at Japan any families' homes conveniently. Hence, it implies that future one day, artificial intelligent machine has possible to be invented to own human's reading and/or writing abilities.

For example,online book publishing is one kind of popular internet technology. For example, Amazon publish is as a business model with many potential advantages, relative to a physical operation. It held out the potential of lower book inventing and distribution costs and reduced overhead. Consumers could find the books, they were looking for more easily and a variety book topic choices could be offered for sale. It can accept and fulfill orders from almost any domestic location with equal ease. And most purchasers made on its site would be exempt from sales tax. One Amazon strategy hand, it would have to make its returns and redress processes transparent and reliable, and offer other ways for clients to learn, as much about the book possible before buying. Future online book market development trend, such as Amazon, Barnes & Noble etc. online book shops.

Hence, online book store technology can be applied to artificial intelligent technology. Such as artificial intelligent machine men can apply computer technology to learn the abilities of reading and/or writing any books either on paper or on computer. Hence, it is possible that artificial intelligent machine men will have similar human's writing and/or reading books ability when they own human's mind ability. However, it bring this questions: Can artificial intelligent machine men own human's mind abilities? If they own human's mind abilities, is it mean that they can write and/or read any books? Can artificial intelligent machine men own human's mind abilities to create to write any books? Can artificial intelligent machine men own human's mind abilities to read and make any judgements or decisions more accurate than human's judgements or decisions? To answer these questions? I shall indicate that online book reading and writing technology can be applied to artificial intelligent machine men reading and writing technology. Because they are similiar computer mind technological development. So, I believe that future artificial intelligence machine men can be invented to own similar human's reading and writing's mind abilities in future one day.

I believe artificial intelligence and online technological reading abilities are very similiar. Nowadays, computer can be invented to attempt to read and write any books by human. Why can not artificial intelligent machine men replace computer to read and write any books? Artificial intelligent machine men can replace human to attempt to write or/and read books, due to artificial intelligent machine men had invented to own human mind to do some jobs and their mind had been invented to be similiar to human behavioral abilities to do these behaviors, e.g. cooking, driving, playing games, singing songs, speaking, listening, frighting etc. different human's abilities. So, it seems that artificial intelligent will be possible to be invented to own human's mind abilities to do any writing or reading behaviors or functions.

## 1.2 Prediction of artificial intelligence reading and writing abilities

development

What is future trend of artificial intelligence reading and writing abilities development? To answer this question, we need to know what benefits of artificial intelligent machine men can attribute to human's needs when they can own any human's mind to read or/and write any books.

I shall indicate e-books reading and writing example, if artificial intelligent machine men can be invented to own human's mind to write and/or read e-books on computer. Then, it brings this question: Can artificial intelligent machine men assist human to learn to do judgement to solve any challenges?

I believe that when artificial intelligent machine men can be invented to own human mind to write or/and read any books, then they will own human's mind ability to make judgement to solve any challenges more accurately, even their decisions can be more accurate to compare to human's decisions. So, artificial intelligent machine mens' writing and reading ability is the main factor to cause their mind to do any judgement in order to make any decisions more accurately. Consequently, in future one day, artificial intelligent machine mens' writing and reading ability will be invented to similar human's reading and writing abilities as well as their minds can also be invented to similar human's minds as well as their judgement abilities can be invented to similar to human's judgement abilities to make any decisions more accurate.

### 1.3 The influences when AI is invented to own human's mind and judgement abilities

Finally, I shall discuss what are the influences when AI is invented to own human's mind and judgement abilities in our future job market. The achievement of artificial intelligent (AI) machine men achievement requirement of owning human's mind and judgement abilities which requires extensive manual labor, and by augmenting the calling process with machine learning, the process where speed and accuracy are needed to close to human's mind and judgement abilities. Expert human race callers now have better information at artificial intelligent machine men at their fingertips faster.

Hence, if the above those requirements are achieved to satisfy artificial intelligent machine men ind and judgement abilities demand to close or exceed humans' mind and judgement abilities. Then, I believe that future human's some simple jobs must be replaced by (AI) machine men. Even, human's some professonal jobs, e.g. lawyer, accountant, administator, typing etc. professional skillful jobs, which will be either replaced or will be assisted by (AI) machine men. For example, (AI) machine men learn how to type english or other language words to do typing job ; they can learn how to apply accounting knowledge to record any firm's income and expenditure record of accounting job; they can also learn how to assist architects to design any architectural building drawing plans to do architect jobs; they can learn how to analyze any court evidences to judge any criminal or civil cases and assist lawyers to give legal advices to achieve more reasonable judgement for any legal cases; they can also learn how to assist firm's managers or administrators to manage any organization teams efficiently.

Consequently, when (AI) machine men can be invented to achieve to exceed human's mind and judgement abilities level. Then, I believe that they can do instead of human' simple jobs, which can do even human's more difficult and more judgement requirement of professional skillful jobs. So, (AI) machine men must need to achieve to do any jobs, they are same, even exceed to human professionals' abilities. Then, it will cause a lot of human's jobs to be disappeared or some human's jobs will be replaced by owning judgement and mind abilities of (AI) machine men to do.

Hence, future many human's jobs will be replaced by technological labors. Employers choose to buy (AI) machine men to replace human labors. The reasons include (AI) machine men have none unhappy, angry emotin to influence their low efficiencies and low productivities. Their judgement and mind abilities can exceed human's abilities or do any jobs to compare better performance to human's abilities. Consequently, different occupation labors need to prepare to learn how to co-operate with (AI) machine men to let future employers feel (AI) machine men will be human's assistant to assist human to do jobs efficiently when human and (AI) machine men work together. It aims to avoid future employers feel (AI) machine men's judgement and mind abilities can exceed any low knowledgeable and skilful occupation labors, even high knowledge and skilful occupation labors. It means that (AI) machine men are only labors' assistant if (AI) machine mens' judgement and mind abilities are below under to human labors' judgement and mind abilities.

Consequently, to avoid (AI) machine men can replace human to do any simple or complex jobs to cause any future any occupation labors' competitiors. I recommend that it is right time labors ought prepare to learn different skills. So, every individual labor does not only concentrate on one kind of skill. Because supposing one kind of the occupation labor's job duties are replaced by (AI) machine men. If the employee had owned more than one kind of occupation skill. Then, I believe that who can avoid the unemployment threat more easier than the employee only owned one kind of occupation skill, when (AI) machine men had invented to own human's mind and judgement abilities in future one day.

Why does AI machine lecturer can raise
education quality

When (AI) machine men can own human reading and writing and judgement and analytical abilities, then they can replace university lecturers to teach students to raise students‘ learning abilities absolutely. Then, it bring this question: Why does I machine lecturer can raise education quality? Why do universities prefer to apply (AI) machine lecturer to teach teachers more than human lectuer in university lecturer hall learning environment? Will (AI) university lecturers replace human lecturers to teach students to learn at university lecturing halls popularly? Can (AI) university lecturers replace university human lecturers to teach students more easily and it can let students feel more easily to learn when they are listening what (AI) university lecturers are teaching to them every time university lecture.

In university today, nearly all students need to attend university lecturing hall to listen their lecturer's teaching in every time course. However, many students do not feel interesting to attend university halls to listen human lecturer's teaching. The reasons include, they are busy, so no time to attend lecturer's hall to listen lecturer's teaching; or they feel bore to listen their lecturer's teaching; they feel difficulty to learn; they have confidence to exam and do their assignments, so they feel that they do not need to go to lecturing halls to listen their human lecturer's teaching. However, if one day, (AI) machine lecturers are invented to teach university students to learn and solve their learning difficulties. Can it raise student individual learning interest, due to (AI) machine lecturers' education quality is better than human lecturers‘ education quality?

What will influence to university students if (AI) machine lecturer can invented to replace human lecture? The influences will include such as below:

First reason: the only way is going to be useful to university lecturers are if all (AI) machine lecturers are well-informed and fully supported to assist human lectuers to teach whose students to let them to listen whose teaching absolutely. So, human lecturers can concentrate on doing any education research and data gathering jobs to prepare for (AI) machine lecturers to help them to explain human lecturers' every time prepared course contents more efficiently. So, (AI) machine lectuers can help human lecturers to share whose teaching time in lecturing halls. Human lecturers' can spend whose hall lecturing time to do whose educational research or other educational gathering jobs absolutely.

The second reason, the human lecturer (Human capital) has ability and efficiency of concentrating on education data gatehering research jobs to prepare to write whose books. When (AI) machine lecturer replace the human lecturer to spend time to attend lecturing hall to teach students. Fo long term, the human lecturer can raise education productivity growth and education quality raising, due to who only concentrate on searching or gathering data to prepare to write whose books to raise their education level.

In macro and micro economic view, the well (AI) machine lecturer educated labor ( human capital) is often replaced to human lecturer as one of the critical factors to influence rapid education productivities and educational quality growth to the Asia developing countries' any regions or cities. Because any of these Asia developing countries, such as China, Korea, Philippines etc. countries which need have well educated and knowledgeable lecturer labors to raise any universities' educational productivities and educational qualities growth. So (AI) assistant lecturer factors which ought have close relationship to cause the good or bad future student learning effectiveness and education or learning qualities raising in these any one of Asia developing countries.

The third reason, for the big population of student growth number example, China's student growth rate is larger than school growth rate. If China expect every students have enough chance to study in schools, but university human lectuer numbers are not enough to supply to universities to teach their students. I believe that (AI) machine lecturer is only one kind of teaching method to solve these big population countries' university lectuer number shortage challenge.

In conclusion, in long term, (AI) machine lecturers can solve university human lecturer shortage challenge as well as they can attract many students to attend lecturing halls and human lecturers can raise education quality when they can concentrate on searching or gathering data to prepare their education career, when (AI) machine lectuers replace them to spend time to attend univesity halls to teach students in every university lecturing time.

Future AI machine education market

I believe that when AI (artificial intelligent machine men) which can invented to own to similar to human mind, learning, language, analytical, judgement abilites. Then, which can be applied to any education market service industy. (AI) potential education market service industy includes such as below:

- (AI) university lecture assistant

Future (AI) machine men can assist univerity lectuers to attend university halls to attempt to teach university students for different subjects, e.g. english, math, economic, math, engineering, art, architect etc. different subjects. It depends on the human lectuter who prepares to spend time to teach the (AI) machine lecturer to remember whose teaching subject. For example, the economic lecturer spend one year time to teach the (AI) machine lecturer to learn all economic knowledge. Then, the (AI) machine lecturer can use its machine brain to remember all the human lecturer's economic concepts and prepared teaching economic contents within the one year. Hence, after one year the (AI) machine lecturer can remember all the human lecturer's economic concepts and economic theories and economic contents to prepare to attend university lecturing halls to teach all first year undergraduated first year economic students confidently. It means that the human lecturer's job duties will change to teach (AI) machine lecturer to learn whose economic knowledge to prepare to let the (AI) machine lecturer to replace whom to teach whose university students; so the human lectuer can spend more time to do other research job for whose university education development. Hence, the (AI) machine lecturer can share the human lecturer teaching job as well as the human lectuer can concentrate on spending time to do whose research jobs for whose university education development. This is one both win strategy to university and the lecturer if (AI) machine lecturer is invented to assist future university lecturer's teaching jobs.

- (AI) secondary and primary teacher assistant

In the future (AI) technolgical development, instead of (AI) machine men can be applied to university education aspect. Future (AI) machine men can also be applied to secondary and primary teaching aspect. For example, primary and secondary schools do not need attend classroom to teach students. (AI) machine teachers can replace them to attempt to do teaching job. They only need to spend one year time to prepare to teach (AI) machine teacher to learn how to apply their teaching skill concern their subjects who need to teach to their students, e.g. english language writing and reading and spelling skill, sing song skill, drawing picture skill, calculation skill etc. different studying skill. Then, the (AI) primary or secondary machine teacher can apply the primary or sendary human teacher skills to attempt yo teach whose students. Hence, the primary and secondary human teacher whose duties will change to learn how to teach whose teaching skills to let the (AI) primary or seondary machine lecturer to remember how to apply human skills to teach whose students for different subjects, such as, english writing and reading and spelling language skills, singing songs language skills, math calculation skills etc. Hence, future primary or secondary school teachers who responsibilities will change to learn how to teach (AI) machine teacher teaching skills to prepare to replace them to teach their students in classroom.

- (AI) scientific research assistant

Future (AI) machine men can be applied to science research aspect, instead of school education job. For example, (AI) machine men can be any scientist's assistant, e.g. space scientist, earth or ocean scientist, human or animal behavioral psychological scientist, climate scientist, chemical scientist, drug scientist etc. How can (AI) machine men can be any kind of scientist to assist scientists to do research jobs ? I shall indiate such as below:
For space science example, the (AI) machine space scientist can assist human space scientist to gather space data to assist space scientist to research any undiscovered material to cause our earth, even space. Hence, the space scientist only need to teach the (AI) machine scientist to learn how to help them to apply space technological tools to gather data and then enter all data to computer to record, even the (AI) machine scientist can store all space data discovered record to their machine brain every day. Hence, the human space scientist does not need to spend much time to do gathering data job. The (AI) machine space scientist can help whom to do these space data gathering job, then the human space scientist can concentrate on spendin time to do space research job in whose space science laboratory every day.
For earth science example, the (AI) machine earth scientist can help the earth scientist to go to anywhere to gather earth or ocean natural activity data in our earth every day. Then, the earth scientist only need sit in whose earth laboratory to wait the (AI) machine earth scientist to come to whose laboratory to give whose gathering every day earth or ocean natural activity data to do future research job. Hence, the earth scientist does not need to

leave whose laboratory to do any data gathing jobs concern earth or ocean natural activities. The (AI) machine earth or ocean scientist had helped him/her to go to our earth or ocean anywhere to do any earth or ocean activities data gathering jobs every day. Hence, the earth or ocean scientist can concentrate on spending whose time to do any research jobs in laboratory. It means that the (AI) machine earch or ocean scientist had replaced whom to do all outdoor original gathering data jobs.

For these human or animal behavioral psychological scientist, climate scientist, chemical or drug scientist, scientist all examples, the (AI) machine human or animal behavioral psychological scientist can help them to do any data gathering job, e.g. the (AI) machine scientist can learn how to help human or animal behavioral psychological scientist to contact human or animal to observe their daily activities and record all their activities data to transfer all these daily activites data to let the human or animal behavioral psychological scientist to do psychological researching analysis only. The (AI) machine climate scientist can help the human climate scientist to arrive anywhere to observe climate changes and record climate changes daily. Then, the human climate scientist only need to wait the (AI) machine climate scientist's gathering climate change data record from whose machine brain to do climate changing predict research job in climate laboratory every day. The (AI) chemical or drug machine scientist can help the drug or chemical scientist to gather data of new drug or chemical from internet channel every day. So, the human chemical or drug scientist only need to do researching job after the (AI) machine scientist transfers all daily chemical or drug information to let them to know from internet channel. It means that the chemical or drug human scientist does not need to spend much time to gather drug or chemical new data development trend from internet. The (AI) machine chemical or drug scientist had helped them to do data gathering job every day.

Consequently, future (AI) machine men can do education and research aspects of jobs duties and their role are only human scientists or primary or secondary teachers or university lecturers whose assistants either to share scientist's data gathering job or share teachers or lecturers' teaching job.

## Chapter 4
## How Robots Change Job nature

Although, (AI) technology will be popular to applied to different jobs, but it still needs social acceptance to replace some human jobs. Today, it is increasingly common for people to use robots in various situations at home and in retail stores, hotels and hospitals. Robots are classified into several types based on their functionality ( service and utility robots or those designed to communicate with humans ) and appearance ( humanoid robots or mechanical robots). The types of robot to which every country attaches particular important in the advance of robotics, reflects the sense of values and preferences of its population . Thus, (AI) will be applied to replace human to do these above different kinds of job nature. For example, U.S. has the highest level of robot utilization at home and an retail stores with its people being the most enthusiastic about the future use of robots. Otherwise, Germany shows a strong tendency to consider robots for industrial purposes, and its people feel strong to the presence of robots in their households. Japanese accepts to apply" human aid robot" that can communicate with humans and they have a high level of familiarity with robots.

Hence, it implied those three countries have accept (AI) to replace human to do any these kinds of job duty and it will influence these three countries' workers lose their old occupations and who will unemployed absolutely, due to many (AI) robots replace them to do their job duties in the future. Also, US will have many retail service workers or retail warehouse workers are unemployed. Germany will have many manufacturing industry's workers are unemployed. Japanese will have many communication industry workers are unemployed, such as telephone service, shopping center services etc. different kind of service industry's service staffs . It will cause these kind of workers' competitive abilities are lost in themselves countries' jobs that require such skills include software developers, court judges, nurses, high school teachers, dentists and university lecturers, these occupations are still difficult to be replaced by (AI) robots.

Are robots taking our jobs or making them? In fact, our societies will have unemployment challenges, even (AI) technology has not created before. However, after (AI) robots invention, some of human jobs will be replaced and it can raise many low skillful and low knowledge level worker unemployment number. However, I think that high productivity driven by increasingly powerful IT -enabled machines is the causes of global labor market problems and accelerating technological change will only make those problems worse.

IT technology brings this question: Are robots killing human's jobs or benefiting human's jobs? I suppose that there is a limited amount of labor to be done. The implication is that technology can create unemployment by displacing workers, such as (AI) invention, because the more efficiently worker work ( using machines or (AI) robots), the loss work there is for workers to do. Even, any new jobs will be better done by machines or (AI) robots, and unemployment will still skyrocket. How do we know that humans will always be better at some work, or more importantly, enough work, than machines or (AI) robots, e.g. human drivers drive more safe or careful to compare (AI) robot drivers. But, the challenge is that it is not ensure that (AI) robots drivers must not drive careless to cause the chance of accident occurrences more than human drivers. However, technological change can be beneficial to innovation, automation and increasing productivity for businesses.

Consequently , it may seem machines can hurt wages and job for low skillful, less educated workers. Also, high educated workers are likely as less educated workers to find themselves displaced and devalued, and more education may create as many problems as it solves. Thus, in negative influence, automation effects on particular jobs shift workers to other jobs that are equally or more desirable. Workers may be highly compensated for possessing human capital that is specialized to a labor market. If technological advance is very rapid, such as (AI) invention, causing a large and very rapid drop in demand in a large labor market, the economy may not be able to absorb the sudden surplus of labor in a short period of timer when (AI) robots are popular to replace some workers to do some occupations in global societies.

For example, self-driving vehicles threaten to send truck drivers to the unemployment office. Computer programs can now write journalistic accounts of sporting events and stock price movement. There are even computers that can grade essay revolutionize some part of teaching jobs. Hence, (AI) robots will have possible to replace human brain to do any judgement, argument, and mind job duties. It implies some occupations which need human' mind will be threaten by (AI) robots, e.g. author, accountant, nurse, engineer. Thus, (AI) robots will have possible to replace some professional and high educated workers' jobs in the future.

But, technology can create new nature of jobs in possible. For example, a 60 minutes program indicated technology is putting new categories of jobs in the sites ( sic) of automation, the 60% of the workforce that makes its living gathering and analyzing information. Also, recession: technology kills middle -class jobs that overall technology is eliminating for more jobs than it is creating by (AI) technology. Hence, human's brain work may be assisted by 60% of (AI) gathering and analyzing information for some occupation , e.g. space scientists, ocean scientists, earth scientists etc.

However, I believe the (AI) invention and human job competition may influence global productivity change. Productivity is economic output per unit of input, the unit of on input can be labor hours( labor productivity), but if (AI) robots replace human job, then the unit of input may be (AI) machine hours ( AI) robot productivity or all production factors including labors, machines and energy ( total factor of productivity). Producing more output with less input can take several forms.

The traditional notion of productivity is a form reorganizing production and/or using better or more technology to produce more output per worker hour. But when (AI) robots invention, the form can be reorganizing production and/or using better or more (AI) robots to produce more output per (AI) robot hour. Hence, if the firm apply (AI) robots to produce its products. Then , productivity improvements in the firm may result in less workers employment, due to (AI) robots replace more worker number to achieve more productivity improvement, it has economic benefits ( less factor of production) , but more production in long term.

Thus, (AI) robots can help any firm to achieve productivity improvement in long term, for example, if unproductve farmers move to the city and start working for high-tech. manufacturers. The shift effect can be more dynamic and disruptive as low-productivity industries lose out in the marketplace to high -productivity industries and the compositional mix of the economy changes. Thus, in the long term (AI) robots can also be beneficial to high productivity industries to bring the mix of economy positive changes.

Moreover, automation will also produce some new jobs in firms that sell the new robot or other labor-saving technology. This means that, in general, there will be shift in the economy in the direction of higher-skill and higher wage jobs. Even if the (AI) robot invention country, US becomes a leader in (AI) robots producing productivity-enhancing technology, it will experience a growth in jobs serving foreign (AI) robots product buyers. Hence, (AI) robots can also create (AI) salespeople, (AI) manufacturing workers , (AI) inventors, scientists, (AI) software designer etc. occupations, when if all society does is move workers from insurance firms, restaurants and car factories to robot factories, productivity will have remained the same to create job needs for insurance, restaurant and car manufacturing worker service occupations for (AI) software designer, (AI) service robots manufacturer, (AI) service robot seller etc. related (AI) service robot product occupation created in (AI) robot technology job market. Hence, (AI) invention also create new (AI) technology job chance. (AI) impacts management job market.

In future, organization management will be changed from (AI) introduction. Division of labor will change and collaboration among humans and machines will increase. Companies will have to adapt their training, performance and talent acquisition strategies to account for a new found emphasis on work that hinges on human
judgement and skills, including experimentation and colloboration.

How (AI) impacts any organizational administrative management work? (AI) 's greatest impact will be on administrative coordination and control tasks, such as scheduling, resource allocation and reporting, (AI)-driven will place a higher premium on what we call " judgement work", the application of human experience and expertise to critical business decisions and practices when information available is insufficient to suggest a successful course of action. This kind of work will require new skills and mindsets; replacing people with machines is not goal in itself. When, artificial intelligence enables cost-cutting automation of routine work, it also empowers value -adding augmentation of human capabilities.

Thus, administrative and routine tasks, such as scheduling, allocation of resources, and reporting, will within intelligent machines, responsibilities that have long been reserved for humans. For instance, a typical store manager or a lead nurse at a nursing home must constantly juggle shift schedules, accounting for staff members' absense owing to illness, vaction, time or sudden departures. Many of these tasks will be automated by (AI). Imagine (AI) writing management monthly reports, it is not a distant dream. Leading news providers and Wall street banks are now using (AI) report generators to write news and analytical reports by drawing on quantitative data. The associated press, for example, expanded its quarterly earnings reporting from approximately 300 companies to nearly 3,000 with the help of (AI) powered software robots, freeing up journalists to conduct more investigative and interpretive reporting. For another example, Jobalime, a job-placement site, uses intelligent voile analysis algorithms to evaluate job applicants. The algorithm assesses paralinguistic elements of speech, such as tone and inflection, products which emotions a specific voice will elicit, and identifies the type of work at which an applicant will likely excel. In the future , (AI) machines can be applied to assist some kind of office administrative jobs duties. It's attractive to office managers to achieve more accurate judgment to do any administrative matters when who can be assisted from (AI) machines. Thus, managers need to spend time to learn how to apply (AI) machine to assist them to do more accurate judgement, and better informed choices. (AI) robots can be applied to improve the speed quality and cost of available products and services, instead of applying on productivity improvement and administrative improvement aspects. Thus, they may also displace large numbers of workers. This, possibility challenges the traditional benefits model of trying health care and retirement savings to jobs.

In an economy that employs dramatically fewer workers to deliver benefits to displaced workers. For example, the worldwide number of industrial robots has increased rapidly over the past few years. The fall prices of robots, which can operate all day without interruption, make them cost- competitive with human workers. In special consideration, in the service sector, computer algorithums can execute stock trades in a fraction of a second, much faster than any human. As those technologies become cheaper, more capable, and more widespread, they will find even more applicants in an economy.

Consequently, (AI) technology brings unemployed number increasing many businesses continued automating their operations rather than hiring additional workers. A trend among technology companies that receive massive valuations with relatively few workers. For example, in 2014 year Google was valued at $370 billion with only 55,000 employees, a tenth the size of AT & T's workforce in the 1960 year. Hence, if automation technologies like robots and artificial intelligence make jobs less secure in the future, there needs to be a way to deliver benefits outside of employment " flexi security" or flexible security is one idea for providing healthcare, education and housing assistance whether or not someone is formally employed.

In conclusion, (AI) and robots technology will raise unemployment to some occupations when (AI) replaces same industries' workers job duties in our societies in the future, but it also create new jobs to raise employment in any related (AI) robots and automated machine products in (AI) manufacturing. (AI) design, (AI) sale self-related industry, when (AI) replaces same industries' workers' job duties.

(AI) journalism, media publishing, digital

communication technology trend

How to apply (AI) technology in digital communication journalism media, publishing industry? Some scientists indicate future (AI) and digital technology may consist such as: voice driven assistants, emerge. For example, Amazon e book publish applying digital technology and (AI) auto printing technology to sell e books to let readers to listen any e book content by (AI) voice driven speaker when they turn on computer to read e book contents; capable phones start to unlock the possibilities of 3D image of mobile story telling. New smart wearables include ear buds that handle instant translation and glasses that talk and hear. China and India will become a key focus for digital growth with innovations around payment online identity, and artificial intelligence. Thus, future (AI) technology can be applied to 3D image mobile story telling, online payment method to dealt online transaction publishing industry.

Thus, future (AI) technology can be applied to online e book publishing industry to make sound books to let readers feel more attractive . Such as Amazon publish has published sound e books to attract readers to choose to read any its books from online. Also, (AI) technology can also be applied to communication industry. For example, some online pure-play news, opinion and entertainment websites. It is a digital communication media, e.g. online journalism blog (AI) technology can be applied to visual storytellers to let online book readers to enjoy to listen to watch and send any online electronic book contents more attractive. Thus, future (AI) technology will be popular to assist any electronic book publishers to publish visual and sound talking storybook to let readers who can watch motive image and listen and read words from e books more attractive.

Thus, (AI) technology can be applied to internet ecommerce publishing or media industry to help any electronic book publishers to publish sound, image motion electronic book to attract global readers to read, even (AI) technology can be applied to digital entertainment industry, e.g. electronic 3D image virtual video games, computer games. It can be also applied to education industry, e.g. the first true digital native generation and are the native speakers of the digital language of computers to let student to learn different languages or translate words to compare to classroom learning more easily. It can be also applied to communication industry, e.g. (AI) mobile phone. Hence, it seems (AI) technology can be applied to publishing, communication, education , entertainment etc. different industries in the future. (AI) technology will be one kind of tool to satisfy human's daily life needs in the future and these industries has one characteristics is that they need to apply internet to operate to operate to do online business.

Thus, it has three trends of (AI) technology and internet technology need to be linked to achieve one kind of attractive technological business to satisfy client's needs. These three trends as below: All consumer trends involve the internet. It will be many consumer's online habits, shopping, working, socializing, watching TV, studying, travelling, listening. Thus, (AI) music, eating and exercising are just a few examples. This is happening because human usually use mobile broadband or Wi-Fi, rather than cables. Thus, (AI) technology will be applied to mobile to satisfy client's need absolutely.

The mobile phone can be more popular to be used more than computer or laptop tools. The reasons are because women dive the smartphone market by defining mass-market use. But as the speed of technology adoption increases mass market use becomes much quicker then before. Successful new technological products and services , such a (AI) mobile phone products now reach the mass market in popular use. It means that the time period when early adopters influence others is shorter than before. Also, since new products and services increasingly use the internet mass markets are not only faster , but are also more important than ever to consumer themselves. Most internet services become more valuable to individuals when many use them. Thus, it causes why (AI) mobile phone will be popular to be used.

Since, new products and services increasingly use the internet, in the future several trends focus on (AI) smart phone users. Consumers' familiarity with using smartphone apps. Essentially, the technologies will bring other related (AI) and internet service needs, e.g. sound and image emotion e book needs, (AI) mobile communication needs, e-virtual games or e-3D image virtual games etc. entertainment activities needs with such a large part of the world's population now online, it is clear that there is strength in numbers.

Thus, (AI) imagines , if future any (AI) and internet related services or products new technology is easy to use and inexpensive, when the latest products reach the mass market almost as quickly as they reach the early adopters and industry experts. I believe that any (AI) and internet related products or services must be popular to accept to consume for entertainment or useful aim. For example, with major players including Apply, Facebook and Google had invested (AI) technology to develop their businesses. (AI) technology has the potential to disrupt everything in the coming years, from the lives of connected consumers to every industry (AI) will be an alternative route for brands to reach consumers with convincing and relevant messages. Digital technology will assist of the future, then it can improve technology to bring this effect, such as sophisticated software machine learning and speech recognition effective. Hence, Google, Facebook , Yahoo web site service companies can apply (AI) technology to help other companies to advertise their businesses, such as travel, retail, and education etc. industries more attractive. (AI) technology can be applied to internet company to be aware and familiar enough to drive among mainstream consumers, it can create online experience to travel, retail , education and other entertainment needs to online consumers to seek their entertainment needs more easily. Hence, in the future (AI) technology and internet related entertainment service needs will be raised in this (AI) and online consumption market.

- (AI) healthcare service industry development

In the future, (AI) medical internet technology tool can be applied to assist individual's health at the center of their focus, e.g. smartwatch compatible mobile app. patients can let personalized reminders for taking their medication snap pictures of their prescriptions to expedite refills, and scan their insurance card. So that, store clerks are prepared with up-to-date patients' information . (AI) owned health operated technological clinics can help patients to receive treatment for minor illnesses, flu shots, cholesterol screenings and more than a dozen other medical services, all of which can be patients who can't make it to a physical location. (AI) healthcare services organizations can provide various telemedicine services. So, patients can receive care via phone or video chat.

For example, one London-based intelligent Brewing company has developed an (AI) system to continuously collect and incorporate customer feedback, which the system itself uses to brew ne various of the company's beers. Thus, the beer clients can give feedback to talk to the algorithm (AI) machine, whenever or anywhere who're drinking the beer. It is such any healthcare services organizations can apply (AI) machine to collect patient's feedback to talk to the algorithum (AI) machine whenever or anywhere who're eating any medicines. So doctors can know every patient's health conditions any time. If the patients feel uncomfortable, the doctor can know from (AI) machine notification to decide whether the patient needs to eat another new medicine or keep to eat same medicine is better. Hence, (AI) medial internet technological body check report machine will be proper to be needed to serve any hospitals' patients in the future.

However , it brings this question. How can (AI) medical internet technological body check report machine apply to hospital more efficient? The essential new medicine co-workers for the health service digital age health service leaders need apply (AI) medical report machines and artificial intelligence to the newest recruits to the workforce bringing new skills to help health service staffs do new jobs and reinventing what's possible, building the health service workforce for today's digital health service demands for patients. Thus, technology-driven health service model innovation from the health service organization outside in and providing digital health service ecosystems for patients to use the (AI) health service equipment will be popular to be accepted to be used.

● I Robot and internet things future machine
men invention

Nowadays, there are some company, which apply internet and (AI) I Robot technology to do any similar human job nature. For fishing industry example, one company, known for creating the Roomba, I Robot is now working with marine conservationists to launch an ocean-patrolling intelligent robot to hunt and manage invasive species, protecting native populations. And evolved industries like precision agriculture are ramping of our increasing population. Area of practice that once seemed impossible to digitize are fundamentally changing because of the impacts of (AI), internet of things capabilities and big data analytics, which have many potentially positive impactions for society.

For textile industry example, automation is nothing new, it has shaped the workplace to replace human jobs to boost productivity in the textile industry. Textile machines have had a generally positive impact over gears, creating value and allowing textile workers to take up more rewarding age will likely continue to create opportunities and lead to new textile industries, companies and textile occupations. It may also compensate for a demographically driven slowdown in the growth of the textile workforce. The future impact of textile (AI) and automatic and internet link is somewhat uncertain. It seems textile industry will be trend to accept (AI) textile workers and internet of thing to replace traditional manual textile workers to produce any shirts, cloths etc. wearing products in factories popularly, during the (AI) textile machine and internet thing technology can be invented to reach the mature stage in the future.

For factory worker transportation job example, they have also expanded their influence, migrating from the factory floor to the service sector and taking the place of humans in a range of activities from financial transactions to transport route optimization. Further (AI) machines and robots are increasingly programmed to learn, meaning they improve with time and undertake cognitive activities. Hence, (AI) machines and internet technology enable automation of work activities to raise factory workers' efficient and performances, also factories can reduce manual worker numbers, due to (AI) machine workers' assistance.

Future, (AI) robotics technologies and internet technique have these different kinds of characteristics: For soft robotics example, it is non-rigid robots construct with soft and deformable materials that can manipulate items of varying size, shape and weight with a single device. For swarm robotics, it coordinated multi-robot systems often involving large numbers of mostly physical robots. For touch/factile robotic example, it robotic body pails (often biologically inspired hands) with capability to sense, touch , dexterity robots example, serpentine robots with many internal degrees of freedom to threat through tightly packed spaces for humanoid robots example, robots physical is similar to human being often bi-pedal that investigate variety capable of performing human tasks , including movement across terrains, object recognition, speech sensing etc. For autonomous cars and trucks example, it is capable of operating with a human pilot, e.g. the unarmed general atomics Predator XPUAV with roughly half the wingspan of a Boeing 737 can fly autonomously for up to 35 hours from take-off to landing, for unmanned aerial vehicles example, flying vehicles capable of operating without a human pilot, the unarmed general atomics predator -XPUAV , with roughly half the wingspan of a Boing 737, and fly autonomously for up to 35 hours from take off to landing, for (AI) chat bots example, (AI) systems designed to simulate conversation with human users, particularly those integrated into massaging apps.

In Dec. 2015. the general service administration of the US Govt. described how it used a chat bot named Mrs. Landingham (a character from the television show the west wing) to help onboard new employees. Finally, for robotic process automation example, class of software robots that replicates the actions of a human being interacting with the user interfaces of both software systems. Enables the automation of many back-office work flows without requiring expensive IT integration . Hence, future (AI) robot machine men will have different functions to be applied to different industries to use in possible.

Statistics Denmark shows that (AI) automation potential robots will influence few jobs are completely automatable , but close to half consists of 40% automatable tasks: It showed example occupations include share of automated, such as brewing machine operators are more than 80%, logging equipment operators are more than 50%, roofers , stock tasks clerks, travel agent are more than 50%, farmers , nursing assistants are more than 30%, physicians, teachers , managers are more than 10%.

For example, humans perform a wide variety of tasks from planting corn to examine spreadsheets, meeting clients and lifting crates in a store. Each of these actions requires a combination of innate or acquired capabilities, internet technique assistance, ranging from social perceptiveness to fine motor skills and natural language understanding. To understand and map automation feasibility by existing technology. Mc Kinsey has developed a framework of 18 technical capabilities that can substitute tasks performed by humans. The capabilities are grouped in five categories: sensory, cognitive, language, social and emotional and physical. So, it seems (AI) robot machine men and internet technique will have possible combination to invent to own human's emotion , language, learning, task skill abilities.

Mckinsey global institute analysis also showed current technologies have achieved different levels of human performance across 18 capabilities include: sensory perception, autonomously infer and integrate complex input using sensors, cognitive capabilities reorganizing known patterns/ categories supervised learnings, generating novel, logical reasoning/ problem solving, optimization and planning, creative, information retrieval, coordination with multiple agents, output articulation/presentation, national language processing, social and emotional capabilities-natural language understanding, social and emotion sense, reasoning output, physical capabilities-fine motor skills, navigation mobility. Hence, it seems (AI) robots and internet technological will combine to invent to own human' some skills to replace human to do some kind of tasks in possible.

In conclusion, future (AI) robot and internet will be needed to link to cooperate together to raise human's work efficiency in popular.

How artificial intelligence replaces human job possibility

What is the risk of automation for jobs to replace human job? In recent years, there has been a revival of concerns that automation and digitalization night after all result in jobless future. As I argue, this might lead to an overestimation of job (AI) automate , as occupations labelled as high-risk occupations often still contain a substantial share of tasks that are hard to automate.

For example, when the share of (AI) automatable jobs is 6% in Korea, the corresponding share is 12% in Australia. Differences between countries may reflect general differences in workplace organization, differences in previous investments into (AI) automation technologies as well as differences in the education of workers across countries. I also discover that (AI) automation and digitalization are unlikely to destroy large numbers of jobs. But, however, low qualified labors are likely to raise costs as the (AI) automate of their jobs is higher compared to highly qualified workers.

In fact, (AI) technology will influence some new technology to replace some human's job, such as driverless car, the largely autonomous smart factory , service robots or 3D printing. These technologies are driven by advances in computing power, robotics and artificial intelligence and ultimately redefine what type of human capabilities machines are able to do.

Hence, question brings whether (AI) invention will influence general human jobs to be replaced by (AI) autonomous jobs? Whether will the potential foe automation with actual employment loss? In particular, the technical possibility to use (AI) machines rather tasks need not mean that the substitution of humans by machines actually takes place.

Whether (AI) technology replaces human's some job, it is beneficial to our society or not. Instead, machines are increasingly capable of performing non-routine cognitive tasks, such as driving or legal writing . In particular, advances in the field of machine learning (ML), e.g. computational statistics and visions, data mining, artificial intelligences allow for automating cognitive task, when the use of (ML) in mobile robotics (MR) also allows for automating certain manual tasks. So, it seems, (AI) technology can replace some labor job, e.g. warehouse transportation, even mind's job, e.g. legal writing, driving in possible.

For example, if (AI) automatic non-manual driving can reduce hurt or death risk, it is beneficial to our society, or (AI) automatic robots can more any heavy things ( products) in warehouse safely. Then, it can reduce the warehouse labor's bodies hour risk, it is beneficial to the workers. Even, if (AI) robot can write any legal documents, no any word errors in short time. It is beneficial to the law companies , but it also bring unemployment chance, due to these jobs can be replaced by (AI) robots to do in the future. Hence, it will cause some occupation to be disappeared, due to (AI) robots can do our these kinds of jobs in the future.

Frey & Osborne (2013) reported these kinds of occupations will be replaced by (AI) robots in possible. They include computer, engineering, financial, management, legal , art and medium, community service, education, healthcare practitioners and technical service, sales and related, office and administrative support, farming, fishing and forestry, construction and extraction, installation, maintenance, and repair , production, transportation and material moving. It seems our future some professional occupations will have possible to the replaced by (AI) robots to replace, instead of labor jobs. Hence, (AI) robots technology will have much trend to replace high knowledge or low knowledge skillful labors in the future.

In conclusion, it implies that only using information on task-usage at the individual level leads to significantly lower estimates of jobs " at risk", some workers in occupations with according to high automate nevertheless often perform tasks with are hard to automate. Why can (AI) replace human to do some kinds of jobs? (AI) artificial intelligence refers to the ability of a computer or a computer enable robotic system to process information and produce outcomes in a manner similar to the thought process of humans in learning, decision making and solving problem. By extension, the goal of (AI) systems is to develop systems to capable of tasking complex problems in ways similar to human's logic and reasons who feel in our future. Hence, it means future (AI) robots has effort to replace human to do any jobs in possible.

(AI) directions for future non-manual
control road vehicles market

Future road vehicle products and technologies must meet social, economic and environmental protection and driving safety goals , and satisfying market requirements for mobility, accident reducing, performance, cost desirability. Thus, (AI) auto-non manual control vehicles need to be followed this direction to invent. To satisfy future driver's safety of needs, enhanced vehicle speed desired functional performance of road transportation system, required and desired technological response, including research needs. It is long term up to 20 years vision, for (AI) auto non-manual research. Thus, (AI) auto non manual vehicle manufacturers need often to revise their (AI) vehicles functions to raise to improve their system performance and driving industry driver's needs, e.g. private drivers need or public transportation driver's need or business client's need. Hence, future (AI) transportation will need have individual driving consumer and business driving consumer both targets.

Thus, future (AI) automation manual control vehicles need to deliver high impact technology solutions to meet social , economic and environmental and safe goals. Engine needs to be improved efficiency , performance, drivability, reliability, durability and speed-to-market together with reduced emissions and cost; hybrid, electric and alternatively fuel (AI) non manual control vehicle technology development, leading to new fuel and power systems, such as hydrogen, fuel cells and batteries, which satisfy future social, economic and environmental and safe goals. Software, sensors, electronics and telematics technology development are needed to be lead to improve vehicle performance, control and adaptability, intelligent , mobility and security, structure and materials technology development, leading to improved safety, performance and leading to flexibility with reduced cost and environmental pollution to achieve the (AI) non manual drivers to feel (AI) vehicle performance, auto control and adaptability is better to compare traditional manual driving vehicles.

In fact, in traditional manual driving market, Japan and USA had had over 80% of world car production by six major global groups. In the future, it is possible only USA can dominate (AI) non manual auto driving vehicle manufacturing market if Japan had no effort to manufacture any (AI) auto non manual control vehicles. So, it means that it is only Japan is USA potential (AI) auto non manual control vehicle manufacturing competitors. Also, it means that it is only USA has effort to export (AI) auto non manual control vehicles to global (AI) auto non manual control vehicle market.

Thus, in long term, (AI) non manual control vehicle product market development, USA (AI) vehicle manufacturers will have these requirement to win new technological competition to traditional manual control vehicle. The requirements include: low cost fuel, low carbon, fuel cell and telematics technologies, the technological roadmap function, such as detailed consideration clear provision of other important areas to the drivers. When the (AI) non manual auto drivers are sitting in the non manual control auto vehicle. Although, who does not need to drive, but who need to know how to go to anywhere by electric road map show clearly. So, the driver won't lose direction and he/she can know the (AI) non manual control vehicles is driving to anywhere in any time, even when who is sleeping.

In the future, the (AI) non manual control vehicles need to be invented to satisfy any business , transportation clients' needs, instead of individual clients needs, e.g. cans, trucks, buses, emergency and utility vehicles, trains, trams etc. Hence, technological road mapping is one important tool to help any business, transportation ( AI) non manual control vehicle clients. Technological roadmap is a technique that is used in industry to support strategic planning for (AI) non manual driving vehicles in the future. Electronic road maps generally take the form of multi-layered time based charts, linking technology developments to future (AI) non manual control vehicle market requirements.

Technology road mapping is a flexible technique and the roadmap architecture and process for developing the roadmap most generally be customized to meeting the particular aims. Why technology roadmap will be popular to (AI_ non manual control vehicles. It's advantages include: It is a technology solutions and options that can enable the performance targets to be achieved engine hybrid, electric and alternatively fueled vehicles, software, sensors electronics and telematics, structures and materials design and manufacturing process. It is road transport system performance measures and targets tool, in response to the trends and get (AI) non manual control vehicle drivers to get society, economy, environment protection, low cost driving benefits, also it can help any transportation clients to know how to go to anywhere clearly. Hence, technological road map will be one good tool to assist (AI) non manual control auto vehicle to develop future road driving market.

Reference(source)

Frey & Osborne (2013), The future of employment: How susceptible are jobs to computerization? University of Oxford.

Mckinsey Global Institute Analysis

Statistics Denmark, Global automation impact model, Makinsey analysis

(AI) development second stage

(AI) -driven automation industry development

(AI) -driven automation industry will create wealth and expand economy growth to any countries, but it will be accompanied by changed in the skills that workers need to learn. One of main ways that technology increases productivity is by decreasing the number of labor hours needed to create a unit of output. It implies (AI) technology will influence low educated and low skillful labor number to be decreased ( reduction employment number). In contrast, technological change tended to work in a different direction throughout the nowadays. The advance of computer and the internet raised the relative productivity of higher skilled workers. So, routine-intensive occupations that focused on predictable tasks disappearance, such as switch board, operators, filming checkers, travel agents and assembling line workers etc. were particularly replaced by new technologies.

However, today, it may be challenging to predict exactly which jobs will be most immediately affected by (AI) driven-automation. The reason is because (AI) is not a single technology, but rather a collection of technologies that are felt unevenly through the economy to influence job changing both negatively and positively. In positively view point, (AI) driven-automation will make many workers more productive and increase demand for certain skills. Consequently, new jobs are likely to be directly create in areas , such as the development and supervision of (AI) as well as indirectly created in a range of areas throughout the economy as higher incomes lead to expanded demand. Otherwise, in negatively view point, many traditional human needed ( demand) skillful jobs will be threatened by automation are highly concentrated among lower-paid, lower-skilled and less -educated workers. It means automation will cause pressure on demand for this group, pressure and employment, if (AI) can replace the low skilled and less educated workers' jobs. Thus, (AI) will have negative influence to impact on the labor market.

(AI) capabilities will enable automation of some tasks that have long required human labor. Why can (AI) replace some simple human jobs? For example, advances in robotics are expanding machines' abilities to interact with and sharp the physical world. Combined , (AI) and robotics will give rise to smarter machines that can perform more sophisticated functions than ever before and brings more advantages that humans have exercised. This will permit automation of many tasks now performed by human workers and could change the shape of the labor market and human activity.

## How (AI) influences labor market

Today, it may be challenging to predict exactly which jobs will be most immediately affected by (AI)-driven automation. Because (AI) is not a single technology, but rather a collection of technologies that are applied to

specific tasks.
Some specific predictions are possible based on the current (AI) technology. For example, driving jobs and house cleaning jobs, bank counter service jobs, telephone enquiry service operators. Restaurant cooking jobs, simple accounting record service jobs etc. that require relatively less education to perform. Advancements in computer vision and related technologies have made the feasibility of fully appear more likely, potentially displacing some workers in driving-dominant professions. Seemingly similar robot, for which the operational tasks is less specific of navigating to a specific destination when following a set of given rules and preserving safety.
In the future, the effects of (AI) on the labor market in the decade ahead will continue the trend toward skill-biased change that computerization and communication innovations have driven in recent decades. Thus, some human driving occupation will be disappeared or replaced by (AI) automation driven. For example, bus drivers, light truck or delivery services drivers, heavy and tractor-trailer truck drivers, school drivers, tax drivers, travel bus drivers.
However, (AI) technology could enable some workers to focus time on other job responsibilities, boosting their productivity, and actually raised wage growth among those still holding the reshaped jobs. For example, salespeople, who currently spend a considerable amount of time driving could find themselves able to do other work when a car drives them from place to place, or inspectors and appraisers could fill out paperwork, when their car drives itself. This (AI) -driven technology should make these workers more productive, with (AI) -driven technology serving as a complement, not a substitute. New jobs will also likely be created, both in existing occupations cheaper transportation costs with lower prices and increase demand for products and all the related occupations, such as service and fulfillment, and in new occupations not currently foreseeable.
What kind of jobs will be created by (AI) technology? Predicting future job growth is extremely difficult, due to it depends on technologies or substitute for existing today as well as they may complement or substitute for existing human skills and jobs. However, (AI) will also lead to substantial indirect job creation to the degree it raises productivity and wages, it may also lead to higher consumption that would support additional jobs from high-end draft production to restaurant and retail. The future(AI) " augmented intelligence", the technology's role is as assisting and expanding the productivity of individuals rather than replacing human work. Thus, based

on the biased-technical change framework, demand for labor will likely increase the most in the areas where humans complement (AI) automation technologies. For example, (AI) technology , such as IBM's Watson may improve early detection of some cancers or other illnesses, but a human healthcare professional is needed to work with patients to understand and translate patients' symptoms, inform patients of treatment options, and guide patients through treatment plans. Shipping companies may also partner workers who pick up and deliver products over the last feet with (AI) enabled autonomous vehicles that move workers efficiently from site to site. In such cases, (AI) augments what a human is able to do and allows individuals to either be move effective in their specially task or to operate on a larger scale. Thus, it seems (AI) technology will also create new jobs, raise productivities and workers' efficiencies.

Redefining management in the workforce of artificial intelligence
In the future, due to artificial intelligence influences to some kind of human jobs nature. So, the kind of human jobs of management methods will also need to change to adapt the artificial intelligence technology input to their organizations. It will cause challenges for every executive and manager if who won't have effort to manage their teams how to apply artificial intelligence technology to work efficiently and easily. For example, division of labor will change among humans and machines will increase. Thus, companies will have to adapt their training performance and talent strategies how to emphasize on work that how to make human judgment and skills and experimentation. Thus, (IA)'s greatest impact will be on administrative coordination and control tasks, such as scheduling , resource allocation.

In fact, mangers will encounter this challenges: How to apply human experience and expertise to judge critical business decisions and practices when the information available is insufficient to suggest a successful course of action? Due to this kind of work will require new skills and mindsets. I shall indicate these change management methods to adapt (AI) technology. Such as: administration and routine tasks, scheduling , allocation of resources and reporting will fall within the intelligence machines, responsibilities that have long been reserved for humans. For example, a typical store manager or a lead nurse at a nursing home most constantly arrange shift schedules, accounting for staff members' absences owing to illness, vacation time or sudden departures.

Thus, the managers need to learn how to arrange new division of labor within the organizations after (AI) technology had been implemented to the organization. Artificial intelligence is currently influencing into once considered exclusive to humans: assessing and acting on human emotions and personality traits. The influences to managers need to change their strategies to adapt (AI) technology implements include such as below:

Firstly, managers need to spend the bulk of their time on coordination and control tasks from intelligent system implements. Their time spending on these major three aspects from impact of intelligent system: coordinate and control, solve problems and collaborate and people and community , strategy and innovation three aspects. Thus (AI) will influence managers need to change their judgment method to teach whose teams how to adapt the (AI) system operations in any organizations.

Secondly, (AI) will influence top, middle and low level management needs to change to adapt the (AI) technology operations to any owned (AI) technology organizations in the future. Intelligent machines must be trained in context. Just like humans , on-the-job training is a requirement for such machines because they typically arrive with only very general capabilities. To get the most from (AI), managers at all levels must participate in the instructional experience and in the learning process and provides managers' familiarity with such systems on these aspects, e.g. How the system works and generate advice, how the system has a proven track record , how the system provides convincing explanations , how the system can make simple rule- based decisions.

Thirdly, managers need to learn how to make judgment more accurate (AI) systems assistance. Although (AI) will invariably take on more routine work and even augment human decision-making, it won't judgment work, the application of human experience and expertise to critical business decisions when the information available is insufficient to suggest a successful course of action or reliable enough to suggest an obvious course of action. For a sense of the nature of judgment work, consider big data marketing and sales analytics. Such analytics often provide insights that can inform promotional campaigns, including predicting which promotions will generate desired sales brand further into the future, marketing executives need use judgment, combining analytics with their own and others' insight and experience.

The application of experience and expertise to critical business decisions and practice represents the real value of human judgment. But, when artificial intelligent machines are implemented to any organizations to assist the low, middle and top level management to make any business judgment. These forms of judgment work that managers can gather data interpretation, idea development more absolute from (AI) machine assistance. Thus, why these level management executives need to learn how to apply (AI) machines to help them to make any business judgment more accurate.

How (AI) influences organizational change

Consequently creative and social intelligence will be in even greater demand as (AI) makes in management and the workforce. This development will represent a long term trend in labor markets , one characterized by intensifying demand and reward for social skills with a growing desire for creative capabilities, managers will seek to fashion of ideas and hypotheses from inside and outside of the enterprise to shape solutions to their most pressing business problems. Thus, (AI) will influence overall organizational team members who have chance to participate any decision to make more accurate business judgment.

Many managers mistakenly view judgment work as only an individual discipline, failing to appreciate that it can also involve decide interpersonal and organizational practices. In more complex settings, judgment is typically a collective outcome of individuals' and teams' diverse perspectives, insights and experiences. And often , the resulting choices are better informed than decisions that an individual would have arrived at on his or her own.

Thus, when any organizations apply (AI) technology to assist managers to gather data and ideas to make any judgment. In these cases, organizations can create the conditions for effective collective judgment by establishing structures , such as " shadow advisory boards" that prompt managers and employees to source and synthesize multiple perspectives. Thus, a traditional organization (firm) might freshen its thinking is t put together a shadow advisory board, comprised of young, digital people who can apply (AI) machine assistance to make judgment work more accurate whether related to people development, problem-solving or strategizing and innovating for considerable degrees of creative and social intelligence.

Thus, on the one hand, (AI) technology machine augmentation and automation can give these advantages to human (organization managers) , e.g. developing people and community, solving problems and collaborating, coordinating and controlling work, shaping strategy and leading

innovation. Besides, on the other hand, the next generation managers need have these individual attitude to treat intelligent machines to be as colleagues.
When, judgment is a human skill, intelligent machines can accelerate human learning that supports it, assisting in data -driven simulations, scenarios and search and discovery activities. Focuses on judgment work, some decisions require insight beyond what data can tell them. This is the sweet sport for human judgment, the application of experience and expertise to critical business decisions and practices. Thus, managers will also need to find ways to learn how to use digital (AI) technologies to tap into the knowledge and judgment of partners, customer external stakeholders and role models in other industries after the (AI) machine had been implemented to the organization.

Future works change:
Automation, employment
and productivity

Human future " micro to macro" industry trends will be affected business strategy and public policy by (AI) technology. In the future (AI) technology will influence those six themes: productivity and growth, natural resources, labor markets, the evolution of global financial markets, the economic impact of technology and innovation and urbanization. However, (AI) technology will bring economic benefits of tackling gender inequality, a new global competition, Chinese innovation and digital globalization.
Nowadays, advances in robotics artificial intelligence, and machine learning are in a new age of automation, as machines match or outperform human performance in a development to any countries. For example, automation of activities can enable businesses to improve performance by reducing errors and improving quality and speed, and in some cases achieving outcomes that go beyond human capabilities. For example, some research indicated automation could raise productivity growth globally by 0.8 to 1.4 % annually; more than 2,000 work activities across 800 occupations. When less than 5% of all occupations can be automated using demonstrated technologies about 60% of all occupations have at least 30% of constituent activities that could be automated. Many occupations will change that will be automated away: Activities most susceptible to automation involve physical activities, in highly structured and predictable environments, as well as the collection and processing of data. They are

most prevalent in manufacturing , accommodation and food service and retail trade and include some middle-skill jobs. For example, such as natural language processing is a key factor. Beyond technical feasibility, the cost of technology competition with labor including skills and supply and demand dynamics, performance benefits including and beyond labor cost savings, and social and regulatory acceptance will be affected by (AI) automation technology. Thus, (AI) automation will impact to influence global employment in those aspects as below:

Firstly, assuming that people are displaced by automation will find other employment. The anticipated shift in the activities in the labor force is of a similar order as the long-term shift away from agriculture and decreases in manufacturing share of employment. Both of manufacturing and agriculture industries which would be accompanied by the creation of new types of work not foreseen at the time.

Secondly, for business, the performance benefits of automation are relatively clear. Thus, the businessmen have opportunities for their micro economies to benefits from the productivity growth potential and macro economies to benefit to encourage continued progress and innovation , investment and market incentives. At the same time, employers must innovate policies to help workers and institutions adapt to the impact on employment.

This will likely include rethinking education and training, income support and safety nets , as well as support for those dislocated, when employees need to leave themselves homes to move to other cities to learn new (AI) automation works. Thus, individuals in the workplace will need to engage move comprehensively with machines as part of their everyday activities, and acquire new skills that will be in demand in the new automation age. Consequently , the scale of shifts in the labor force over many decades that automation technologies can be a similar order to the long -term technology -enables shifts in the developed countries' workforces away from agriculture in the 21 th century. Those shifts did not result in long-term mass unemployment because they were accompanied by the creation of new types of work not foreseen at the time. However, human will still be needed in the workforce when the total productivity gains are caused by (AI) technology.

What occupations will be influenced by (AI) technology.

In the future, scientists predict that these occupations will be influenced by (AI) technology mostly. They include : retail salespeople, food and beverage service workers, language or translation teachers, health practitioners. Since these work activities have a more relevant occupations are made up of a range of activities with different potential for (AI) automation . For example, a retail salesperson will spend more time interacting with customers, stocking shelves , or ringing up sales. Each of these activities is distinct and requires different capabilities to perform successfully.

Thus, these job activities have similar simple control characteristics. Simple activities include greet customers, answer questions about products and services, clean and maintain work areas, demonstrate product feature process sales and transactions. All these activities can have similar simple activities in order to (AI) machines can be learn how to do these activities from (AI) technology . For example, the capability perception includes sensory perception, cognitive capabilities, such as retrieving automation, recognizing known patterns( supervised learning), logical reasoning problem solving.

Thus, (AI) machine is such human, which has feeling and emotion, such as social and emotional sensing, judgement reasoning methods, natural language understanding and physical capabilities, such as mobility , navigation, gross motor skill, fine motor skills. It seems that the future, (AI) human invents machines which will have these human characteristics to do human similar behavioral job duties more easily and efficiently. It implies these above human occupations will be replaced by (AI) human invention machines in the future. Due to (AI) creation, it is possible to cause unemployment number of these above workers will increase because (AI) machines can do their similar job behavioral activities.

Consequently, employers won't need to employ many of these skillful labor. Otherwise, they can buy less number (AI) machines to attempt to do whose job activities more easily and efficiently. So, it seems (AI) machines will have more high work performance to replace these occupation workers' work performance. Finally, these occupation worker unemployment number will only increase when the (AI) machines had been invented to achieve to do their work behavioral activities absolutely success in the future.

Whether (A) technology machine labor will replace human worker more or assist human worker more

There is no single agreed definition of a robot how outcome of a task that is completed without human intervention. When some definitions require the task to be completed by a physical machine moves and respond to its environment, other definitions use the term robot in connection with tasks completed by software , without physical embodiment.

However, to answer the question : Whether (AI) technology machine labor will replace human worker more or assist human worker more. I shall indicate some examples to let readers to judge whether (AI) technology can create new jobs or reduce old jobs.

Firstly, I shall explain what (AI) function is. (AI) is a service robot that performs useful tasks for humans or equipment excluding industrial automation application . Thus, the classification of a robot into industrial robot or service robot is done according to its intended application. It is also a personal service robot or a service robot for personal used for a non commercial task, usually by lay persons . Examples are domestic servant robot, and pet exercising robot. It is also a professional service robot or a service robot for professional used for a commercial task, usually operated by a properly trained operator. Examples, are cleaning robot for public places, delivery robot in offices or hospitals, fire-fighting robot, rehabilitation robot and surgery robot in hospitals. Thus, these functions will be future (AI) application to our daily life necessaries or business necessaries.

However, some authors agree (AI) will bring negative outcomes of automation, due to raise competiveness, reduce human job nature. Otherwise, other authors argue (AI) will bring positive outcomes of automation, due to raise productivities, job creation, assist humans work.

On the positive outcome hand, robots can increase productivity . This is particularly important for small-to medium sized businesses both are in developed and developing countries economies. It also enables large companies to increase their competitiveness through faster product development and delivery. Increased use of robot is also enabling companies in high cost countries to re shore, or bring back to their domestic base parts of the supply chain that will have previously outsourced to sources of cheaper labor. Currently , the greater threat to employment is not a automation, but an inability to remain competitive. Automation has led overall to an increase in labor demand and positive impact on wages. The reason is that the middle-income/middle-skilled jobs have reduced as a proportion of overall contribution to employment and earnings leading to fears of increasing income inequality, the skills range within the middle income bracket is large. Thus, robots are driving an increase in demand for workers at the higher -skilled and with a positive impact on wages. This issue is how to enable middle-income earners in the lower-income range to unskilled or retain. Finally, the (AI) positive impact supporter who argue the future will be robots and humans can work together.

However, on the negative outcome hand, robots can substitute labor activities, but don't replace jobs. They believe that less than 10% of jobs are fully automatable. Increasingly , robots are used to complement and augment labor activities, the net impact on jobs and the quality of work is positive. Automation can provide the opportunity for humans to focus on higher-skilled, higher-quality and higher-paid tasks. Robots can improve productivity when they are applied to tasks that which perform more efficiently and to a higher and more consistent level of quality than humans. For example, increased productivity is enabling some firms, such as Whirlpool, Caterpillar and Ford Motors company in the US restructure their supply chains, bringing back parts of the manufacturing process to the country of origin. Thus, productivity gains due to robotics and automation are important not just at the company level, but also for build industry and nation competitiveness.

I suppose that productivity can be raised. What are the impacts of robots on employment? Firstly, the main focus of development has been on personal entertainment, which does not drive worker productivity ( manufacturing production). When the internet ( information and communication technology (ICT)) innovation. This is borne and by findings that manufacturing productivity, which has been driven by innovations in automation rather than consumer technologies, has government strongly than productivity in the services sectors of the economy in most nature economies. It seems (AI) automation will create many jobs in internet communication entertainment game industry. For example, many young people like to use internet to play any electronic games from computer or mobile at home or outside home conveniently. Thus, (AI) automation will increase demand to be invented to any new entertainment game from internet channel. It will need to employ many (AI) entertainment game inventors to create many automation entertainment games. Thus, (AI) automation in internet entertainment game industry will need human (AI) entertainment game inventors to invent the knowledge-based capital of (AI) automation entertainment games. The (AI) entertainment game inventors will need own research and development skills, form specific skills, organizational know-how skills, databased knowledge, design and various forms of intellectual property to do these (AI) automation entertainment game invention occupations in the future.

International Federation Of Robotics(2016) indicated that China will be as a major robotics manufacturer and user of robots, benefiting from jobs created by robot manufacturing and productivity gains from robot use. Chins had sold of robots to any one single market every year since 2017 year. The Chinese government has included a focus on robotics in its 10 year strategy. In order to achieve its target of a robot density of 150 units per 10, 000 workers by 2020 year. Thus, Chinese companies will have to install around 650,000 new industrial robots between 2016 to 2020 year, 2.5 times more than installed globally in 2015 year.

Hence, China (AI) manufacturing industry will need to employ many workers . It implies (AI) manufacturing industry will create many new occupations in China. Also, ministry of economy, trade and industry (2015) also showed that Japan currently has the largest stock of industrial robots in operations, primarily in the automation industry. Driven by a rapidly aging population and low productivity rates, the Japanese government has sights on a 20-fold increase in the use of robots in the non-manufacturing sector and a three-fold growth rate of labor productivity in the service sector both by 2020 year. Thus, it also implies Japan will need many robots to be provide to service industry. Due to robots will provide to serve any businessmen's clients. Thus, it is possible that the service workers won't be dismissed as well as it is depended on the serving job nature to decide whether Japan's service workers can still serve to their employer when the service (AI) robots are applied to whose employers.

Consequently, it seems that (AI) can create employment, Ministry of economy, trade and industry (2015) showed that such as China will develop the major (AI) automation manufacturing industry. The (AI) employers will need to employ many workers to manufacture any these different kinds of (AI) robots to satisfy China or overseas individual or business buyers needs. But, (AI) can also cause unemployment to the low skillful service workers. Such as if Japan some service businesses choose to buy any (AI) service robots to replace their service staffs to serve their clients. It is possible that the service staffs will be dismissed, due to (AI) robots can do such as their same service job duties to achieve better service performance.

Thus, today, it is increasingly common for people to use robots in various situations at home and in retail stores, hotels and hospitals these service industries. Robots are classified into server types based on their functionality ( service and utility robots or those designed to communicate with humans) and appearance ( humanoid robots or mechanical robots). The type of robot, to which each country allocated particular importance in the advance of robotics, reflects the sense of values and preferences of its population. Thus, if the country has high population needs to use robots, then they will influence either more new jobs creation or more old job loss in the country's (AI) manufacturing or (AI) service industries both. For example, Japan respondents often associate the term " robot " with humanoid robots that can communicate with human and they have a high level of familiarity with robot. The US has the highest level of robot utilization at home and in retail stores with its people being the most enthusiastic about the future use of robots. Germany shows a strong tendency to consider robots for industrial purposes and its people feel strong effort to the presence of robots in their households.

In conclusion, to judge whether how (AI) will influence the country's employment to be better or worse. It will depend on the country home buyers (users) or business buyers (users) how to use (AI) for their daily needs. If the country , such as US retail stores need to use (AI) , it will have possible to reduce some or many retail service workers. Even, if the country , such as Japan has many home users need to use (AI) , it will not influence the employment market. Otherwise, it will raise (AI) salespeople numbers. Even, if the country, such as Germany and China will have many (AI) manufacturers, then it will create many (AI) manufacturing occupations for these (AI) manufactory workers.

Consequently, (AI) robots manufacturing and service needs will have positive or negative impact to any country's employment. It will depend on the (AI) service provision and service workers' job nature as well as the manufacturing workers of (AI) knowledge level to decide their employment chance in their country's employment market.

What does artificial intelligence(AI) development stage mean

- What (AI) function is?

Some scientists explain that artificial intelligence means which is an expert system, computer software that embodies a portion of the specialized knowledge of a human portion in a specific, narrow domain, owns decision making ability of human expert. The (AI)technology is based on the premise that what makes a person an expert is years of experience that enables who recognizes certain patterns in a problem as being similar to pattern. For example, in the future artificial intelligence system can be applied to control air traffic, design to computer configuration, medical diagnosis, instruction/ training, speech/interpretation, monitoring to ( nuclear plant), planning to mission, factory scheduling, prediction weather, repairing telephone, automatic driving etc. different industries.

Artificial intelligence characteristics include: creative, adaptive , common sense, fact processing, quick replication, broad focus permanent and consistent skill. Otherwise, traditional computer expert system disadvantage includes perishable, unpredictable, slow reproduction, expensive, slow reproduction, slow processing lacks inspiration, needs instruction, narrow focus only machine knowledge. So, artificial intelligence is a branch of computer science devoted to creating computer to influence software and hardware to attempt to create human intelligence or human intelligent behavior. It is learning from experience, responds flexibility in situation that are, new or not anticipated.

Thus, (AI) can be learnt programmed knowledge to solve problems, using reasoning in solving problem, understanding and inferring facts and rules, recognizing the relative importance of different elements in a situation. In summary, artificial intelligence is concerned with two basic ideas mainly: The first idea, it involves studying the thought processes of humans to understand what intelligence is; the second idea, it deals with representing thought processes using companies to create artificially intelligent entities for testing the theories of intelligence.

- Can (AI) impact human job nature?

Human need concern this question: Will artificial intelligence (AI) reduce some human jobs in order to instead of replacing machines to do? Due to artificial intelligence is the ability of machines to do thing, that people would require intelligence. For example, artificial intelligence machine man driving( self-driver), it (AI) machine man driving research is an attempt to discover and describe aspects of human intelligence that can be simulated by driving machine functions. Alternatively, (AI) mathematical research may be another viewed as an attempt to develop a mathematical theory function to describe the abilities and actions of things ( natural or man-made) exhibiting intelligent behavior and server as a design of intelligent calculation machine function.

Why do humans need artificial intelligence machines to instead of traditional human service job? For example, can artificial intelligence machine man (self-driving) driver drive to replace human driver? I shall compare the differences between humans and computers : The characteristics of humans are good at recognizing various things, either seen before or not, recognizing the relationship patterns between things. Human thinking is common sense reasoning, combining all types of sensory input, acting appropriately in novel situations, learning new things and changing behavior patterns, making decisions , even when given incomplete information, working with noisy, incomplete information gathering behaviors . However, characteristics of computers are good at: The tasks humans do naturally are extremely difficult for a computer program as intelligent, which must be able to do the same kind of tack as humans do naturally.

Hence, (AI) is an combination of many different success and technologies: Linguistics - computational and socio, philosophy-logic, philosophy of mind and of language, electronical engineering -image and speech processing, pattern recognition, robotics, machine learning, neural networks, optimization scheduling, management information system and decision making. So, it is possible that (AI) can impact human job nature to instead of human working behavior in the future.

- How can human society job nature
to be changed to artificial intelligent society?

From the first intelligent perspective reason view point, artificial intelligence is making machines " intelligent" acting as humans expect people to act. Artificial intelligence has ability to distinguish computer responses from human responses, it owns knowledge to solve expert problem. From another research perspective reason view point, artificial intelligence is the study of how to make computers do things which, at the moment, people do better ( Rich & Knight, 1991, p.3).

(AI) researchers are native in a variety of domains, e.g. formal tasks ( mathematics, games), tasks ( perception, robotics, natural language, common sense reasoning), expert tasks ( financial analysis, medical diagnostics, engineering, scientific analysis and other areas).

From the second business perspective reason view point, (AI) is a set of many powerful tools, and methodologies for using those tools to solve business problems. From a programming perspective reason view point, (AI) includes the study of symbolic programming problem solving and search .

From the third human technological perspective reason view point, today's computer can do many well-defined tasks, for example, arithmetic operations, are much faster and more accurate than human beings.

However, the computers‘ interaction with their environment is not very sophisticated yet. How can human test whether a computer has reached the general intelligence level of a human being? Can a computer convince a human interrogator that it is a human? But before thinking of such advanced kinds of machines, human will start developing our own extremely simple " intelligent" machines.

So, it is possible that human society job nature will to be changed to artificial intelligent society when (AI) technology is developed to the mature stage in the future.

- Why does human need artificial intelligence machines?

One of major division in (AI) is between humans who think (AI) is the only serious way of finding out how we ( human) work and human who want companies to do very smart things, independently of how we ( human) work. This is the important distinction between cognitive scientists vs engineers. One of another major division in (AI) is between symbolic (AI), which represents information through symbols and their relationships. Specific Algorithms are used to process these symbols to solve problems or deduce new knowledge and connectionist. So ( AI) , which represents information in network. Biological processes underlying learning, task performance and problem solving are imitated from human mind behaviors.

Thus, it is possible that artificial intelligence machines can do the better judgicious behavior to compare human.

- How does artificial intelligence influence future working changing in automation employment and productivity aspects?

In the automation changing influence aspect, as companies increasingly use robots on production lines or algorithms to optimize their logistics manage inventory, any carry out other core business functions. Technological advances are creating a new automation age in which ever-smarter and more flexible machines will be deployed on an ever larger scale in the marketplace. However, researching artificial intelligence with how influences human working nature. We need to answer these questions: How will automation transform the workplace? What will the implications for employment? And what is likely to be its impact both on productivity in the global economy and on employment?

Advances in robotics, artificial intelligence, and machine learning are growing in a new age of automation as machines match or outperform human performance in a range of work activities, including ones requiring cognitive capabilities. What factors are determined the changing in

workplace adoption by artificial intelligence innovation? What advantages are automation? Automation of activities can be enabled businesses to improve performance by reducing errors and improving quality and speed, and achieving outcomes that go beyond human capabilities.
Some scientists indicated based on their scenario modeling. They estimated automation could raise producing growth globally by 0.8 to 1.4 percent annually. Almost, the activities people are paid almost $16 trillion in wages to do in global economy have the potential to be automated by adopting currently demonstrated technology. According to their analysis of more than 2,000 work activities across 800 occupations. When less than 5% of all occupations have of least 30% of activities that could be automated. They also indicated that technical economic and social factors will determine automation. Continued technical progress, for example, in areas such as natural language processing is a key factor beyond technical feasibility , the cost of technology, competition with labor including skills, and supply and demand dynamics, performance benefits including and beyond labor cost savings and social and regulatory acceptance will affect ( alter) the scope of automation.
Other some scientists also indicate U.S. country for example, the anticipate shift in the activities in labor force of a similar order of magnitude as the long term sight away from agriculture and decreases in manufacturing. Share of employment in the United States both which were achieved. So, those factors can influence why artificial intelligence technology needs. So, it is possible that future agriculture and manufacturing both industries will apply (AI) technology manufacturer-kind of job nature to raise productivity instead of farmers, fruit picking workers, farming transportation labours as well as factory manufacturing workers and supervisors etc. human-kind of job nature.

- Is artificial intelligence possible to replace labor ?

Not just intelligence, but also debating, if machines are capable of having a conscious minds. Artificial intelligence has those characteristics as below:
On functionalism aspect, artificial intelligence inputs mental states, sensory inputs, ( beliefs, desires being in pain feeling) and behavioral outputs. Since mental states are identified by a functional role, which are thoughts to be manifested in various systems. Even, perhaps computers which are physical devices with electronic substrate that inform computations on inputs to give outputs similar to brains which are artificial intelligence composed of part any intrinsic relationship to each other. Thus, artificial intelligence

activities is not the whole itself, but into parts or on external influence on the parts.

On dualism aspect, artificial intelligence is a set of views about the relationship between mind are matter. On materialism aspect, it builds the only thing that exists is matter, including consciousness.

On biological naturalism aspect, it is similar a human brain than feels pains makes mental situation. So, artificial intelligence is similar biologist which might to be excited to human labor work.

Hence, it seems artificial intelligence can change ( alter) or replace human labor work of nature in possible in the future.

- Can (AI) technology replace human labour nature of work?

On technological innovation reason view point, the history development of artificial intelligence studying the intelligence is one of most ancient scientific discipline. The history development of artificial intelligence what aims to achieve human use to sense, learn remember and think, logic probability, decision making and calculation develop from mathematics, instead of replacement human labor functions.

Artificial intelligence history development aim is the scientific analysis of skills in connection and practice with the appearance of computers from 1950 year beginning. The artificial intelligence (AI) can deal with the ultimate challenges. How can ( either biological or electronic) mind sense, understand and manipulate a world that is much simple and more complex than itself? And what if would human like to construct something with such capabilities?

The general-purpose software of the early period of (AI) were only able to solve simple tasks effectively and failed when which should be used in a wider range or an more difficult tasks. One of the sources of difficulty was that early software had very few or mix knowledge about the problems which handled, and activities successes by simply syntactic manipulation. Moreover, the other difficulty was that many problems that were tried to solve by the (AI) were untreatable.

The early (AI) software whether trying step sequences based on the basic facts about the problem that should be solved, experimented with different combinations till which found a solution. From the end the 1960 year, developing the so-called expert systems were emphasized. These systems had ( sue-based) knowledge base about the field which handled. Till to the beginning of the 1970 year, ( Prolog) the logical programming language was born, which was built in the computation realization of a version of the resolution calculus. ( Prolog) is a remarkably prevalent tool in developing expert systems ( on medical, judiciary and other scopes), but natural language parsers were implemented in this language. Then, in 1981 s, the Japanese announced the fifth generation computer system project, a 10 years plan to build an intelligent computer system that use the ( Prolog) language as a machine code. Nowadays, (AI) can be applied any industries, such as car manufacturing industry can use (AI) technological machine-men manufacture car, instead of replacing human labors in factory. Even, in the future, using (AI) machine-men drivers can drive any private cars or public transportation tools, instead of replacing human drivers, e.g. bus, train, tram, ferry etc. Also in the future, machine-men can replace housewives to serve families to do housekeeping clean job , e.g. cleaning toilets, bathrooms, kitchens, even cooking functions at home. So (AI) machine-man can reduce housewives works at home. Moreover, (AI) machine man can take care old people , when who are living at homes or elder care centers.

So, it seems artificial intelligence (AI) will be possible developed to manufacture a new generation machine-man to assist ( serve) families to do any simply cleaning or cooking jobs at homes. Moreover, the overall demand of ( AI) general social needs will also rise, such as security, driving transportation tools, restaurant cleaning, elder centers care service etc. So, it seems that individual or families or social needs of (AI) will be increase in the future. Thus, it will influence macro economy growth (GDP) if there are large house family consumer group and hotel or bus or taxis or ferry etc. different business consumer group demand any artificial intelligence machine numbers increasing. Then, the artificial intelligence products and material manufacturers must need to buy many artificaial intelligence materials to produce any kinds of artificial intelligence machines to prepare to satisfy consumer individual needs. Consequently, macro economy will grow to the owned artificial intelligence development countries, e.g. US, China, UK.

- Why can artificial intelligence satisfy human needs?

First, On machine-man satisfactory demand aspect view point, it makes computers that think, it is the automation of activities. We associate with human thinking: like decision making, learning. It is the act of creating machine that perform function that require intelligence when performed by people. It is the study of mental faculties through the use of computational models. It is the study of computations that make it possible to perceive, reason and act. It is a branch of computer science that is concerned with the automation of intelligent behavior. It is anything in computing service that human don't yet know how to do property.

Second, on thought aspect artificial intelligence means systems thank think like humans, systems that think rationally.

Third, on behavioral aspect, artificial intelligence systems that act like human and that systems act rationally. However, the basic objective of (AI) is to represent human's thought processes in computation . These machines are supposed to exhibit behavior that. It is performed by a human being, would be considered intelligent. However, some authors feel (AI) has disadvantages, such as it is not creative, it is excited in the use of sensory devices, it can't make use of a very wide context of experiences and it does not use common sense.

For speech recognition and understanding function needs example, (AI) can be applied in speech recognition and understanding function, which (AI) speech or voice recognition is a data input method. For example, the computer recognizes and understands one ( or a few) word commands. Speech understanding on the other hand is the computer's ability to understanding a spoken language. That is , the computer understands the meaning of sentences, an paragraphs through (AI).

So, (AI) can be attempted to learn human language how to speak. It is similar to translate human language skill, instead of actual human speaking skill. Also, (AI) can assist handicap learning or language student how to listen different languages by machine-man sounds from computers more accurately.

So, it seems that it (AI) can replace human language teachers speaking function and can change teaching language nature of job in language speaking and listening education industry.

- Is artificial intelligence one good choice for human future technological benefit?

Nowadays, new technology development is popular. However, artificial intelligence is one kind of new technology choice among different technologies innovation. So it brings this question: Is artificial intelligence technology value to invest? To answer this question. I shall indicate some other new technology developments to compare (AI) technology development to judge which has urgent needs to achieve human expectation nowadays.

For example, why is green peace interested in new technologies? New technologies features prominently in our ongoing campaigns against genetic modified crops and number power. However, which are also an integral part of our solutions to environmental challenges, including renewable energy technologies, such as solar, wind and wave ( water) power energy as well as waste treatment technologies, such as mechanical, biological treatment.

It seems humans need concern how to apply (AI) technology to solve environment pollution challenges in our future. So, environment protective, agriculture, natural energy technology will be popular demand to attempt to apply (AI) technology to solve their challenges or apply (AI) to assist to develop their industry.

How AI development stage to influence economy growth

How can artificial intelligence technology influence economy?

Advances in artificial intelligence (AI) technology and related fields have opened up new markets and new opportunities progress in critical areas, such as health, education, energy, economic development, social welfare and the environment pollution.

(AI) automation will continue to create wealth and expand the global economy development in the future. However, when many will benefits that growth won't be costless and will be accompanied by changes in the skills, that workers need to increase productivity in the economy and structural changes in the economy. So, in the skills that workers need to succeed in the economy and structural changes.

I shall indicate why aggressive policy action will be needed to help Americans who are disadvantaged by these changes , due to (AI) technology is caused. For automation industry change example, artificial intelligence (AI) capabilities will enable automation of some tasks that have long required human labor. These artificial intelligence technology introduction can increase new opportunities for individuals. The economy and society, but (AI) has also the potential to disrupt be current livelihoods of many Americans. However, (AI) leads to unemployment and increase in inequality over the long run depends not only on the (AI) technology itself, but also on the institutions and policies that are changed.

Thus, it is possible that (AI) technology will raise some countries unemployment number if the employer apply (AI) technology workers to work instead of human labor in their factories, but it can also raise productivities for these employers.

- Can (AI) influence global economy growth?

Technological progress is main driver of growth of GDP per capita, allowing output to increase faster than labor and capital . However, technology can increase productivity, but also decrease the number of labor hours needed to create a unit of output. So (AI) causes unequal to labor wage decreases, even reduces the number of labor to manufacture, e.g. artificial intelligence technology of automation car manufacturing industry; clothing manufacturing industry; plane manufacturing etc. high technology of artificial intelligence manufacturing method. But (AI) should be potential environment benefit, although it raises unemployment ratio. Moreover, it can rise production , due to many skilled craft were replaced by the combination of machines and lower-skilled labor. The result of (AI) technology introduction , it causes output per hour risen when inequality declined, driving up average living standards, but the labor of some high-skill workers was no longer as valuable in the market. Otherwise, if (AI) technology is continue developed to be success. Some routine intensive occupations will be loss, which focused on predictable, e.g. easily programmable tasks, such as switchboard operators, filing clerks, travel agents, and assembly line workers would be particularly replaced by new (AI) technology. However, at the same time, (AI) technology development will bring these benefits: improvement in education ( training (AI) technology scientists) , due to (AI) manufacturing technology needs are raising to businesses and institutional changes, such as the reduction in unionization and raising in the minimum wage to the (AI) manufacturing technology skilled labor in factories.

Because (AI) technology is not a single technology, but rather a collection of technologies that are applied to specific tasks, the effects of (AI) will be felt unevenly though the economy. It will bring some tasks will be most easily automated than others , and some jobs will be affected more than others, both negatively and positively. Finally, new jobs are likely to be directly created in areas , such as the development and supervision of (AI) as well as indirectly created in a range areas though out the economy as higher incomes lead to expanded demand.

However, if (AI) technology could dominate global labor markets. If labor productivity increases, do not influence into wage increases, then the large economic gains brought about by (AI) technology could be increased wealth inequality, due to employers can reduce production cost, but workers (

labors) wages will not be increased, even will be decreased. Hence, it seems the (AI) technology will bring disadvantages to labor market to cause unemployment or reduce wages in possible, although it can reduce employer individual salary (wage) expenditure and it can raise productivity.

- How can artificial intelligence impact global economy growth?

Artificial intelligence (AI) technology is a branch of computer science that aims to create intelligent machines that work and react like humans. So, (AI) is a technology that appears to impact ( influence) human preference by learning, understanding complex contents, enhancing humans in executing both routine and non-routine tasks. In the future, (AI) technology that can be virtual personal assistant, as well as it may exist, such as robots with human-like processing capabilities.

How can (AI) technology impact global economy growth over the next 10 years? During this time period, (AI) technology is predicted to have wide-ranging applications including: Machine learning that automates analytical model building by using algorithms that allow machines to operate without human assistance.

In global education aspect, potential applications include predicting cause-and-effect relationships from biological data, identifying new drugs, self-driving cars, and protecting against fraud, improved natural language processing that allows computers to continue to better analysis, understand and generate language to interface with humans using natural human languages. For example, transcribing notes dictated by physicians, automatically drafting articles and translating text and speech. So (AI) technology can be applied to education aspect to improve humans' knowledge level.

In visual art aspect, (AI) machine vision that allows computers to identify objects, scenes and activities in images. Current applications of (AI) machine vision include providing objective descriptions for the blind seeing( visual) needs.

We except the economic effects of (AI) technology to include both direct GDP growth from sectors that develop or manufacture. (AI) technology and indirect GDP growth through increased productivity in existing sectors that employ some form of (AI). If (AI) technology is an increasingly critical component of more products, it will become an integral part of many people's lives. Thus, (AI)'s ability to influence economic activity, rather than the economic or development status of the region. (AI) has the potential to impact income classes and to bring significant gains to both developed and developing countries. For example, (AI) has the potential to optimize good production around the world by analyzing agricultural regions and identifying what is necessary to improve crop yields.

In estimating the future economic effects by (AI) technology innovation, it is important to note that it is challenging to accurately predict which applications of (AI) will ultimately be commercially successful. In micro level economic influence, we need to apply methodologies to estimate the economic effects of investment in firms developing (AI) technology since investment levels in a technology are a telling sign of the future potential of that (AI) technology.

● How can (AI) influence GDP of high income countries in the next ten years?

How (AI)'s development may affect the global economy over the next ten years. In fact, (AI) technology has the potential to affect business across the global in a wide range of industries in ways only a number of technologies have done in the parts. For example, (AI) technology's expected to be a useful tool for enhancing human capabilities and in some instances replacing functions, such as driving a car, adoption of broadband internet, mobile telephone, industrial robotic automation have served to enhance human capabilities.

However, significant public debate has focused on projections of (AI) technology's effect on the labor force. However, large companies prefer to invest in (AI) technological industry. For example, face book's (AI) research lab., google machine intelligence lab. and micro soft machine learning and artificial intelligence research division are all making advances in (AI) technology and investing in the industry's top talent. Additionally, between 2010 year and 2015 year, nearly $5 billion in venture capital funding invested in firms across the global developing and employing (AI) technology ( Facebook (AI) Research).

● How can artificial intelligence impact on workplace?

Modern information technologies and the labor economy growth of machines is powered by artificial intelligence have already strongly influenced the world of work in the 21 ST century. Computers, algorithms and software simplify every tasks and it is impossible to image how most of our life could be managed without them. How can be the information economy characterized by exponential growth replaces the most production industry based on economy of scales? What will the future world of work look like and how long will it take to get? Will the future world of work be a world where humans spend less time earning their livelihood? Alternatively, are mass unemployment, mass poverty and social distortions also possible scenario for the future, where robots, artificial intelligence systems play an increasingly central role? These questions concern how artificial intelligence further development . Can influence labor economy growth on workplace ? When the labor market has widespread impact on intelligence property, information technology, product liability, competition and labor and employment laws.

How (AI) technology impacts on labor workplace.

The future influence any organizations how labor economies use of (AI) can be analyzed, such as deep machine learning is based on a set of model high level data. Unlike human workers, the machines are connected the whole time in workplace. If one machine makes a mistake, all autonomous systems will keep this in mind and will avoid the same mistake the next time.

Over the long run intelligent machines will win against every human expert. Production robots have been replacing employees because of the (AI) technology. They work more precisely than humans and cost loss. Creative solutions like 3D printers and the self learning ability of these production robots will replace human workers, the automatic data recording and data processing, traditional back office activities are no longer in demand. Autonomous software will collect necessary information and will send it to the employee who needs it. Additionally, dematerialization leads to the phenomenon that traditional physical products are becoming software. For example, CD or DVDs are being replaced by streaming services. The replacement of traditional event ticket, e-travel ticket service products or hard cash will be the next step, due to the possibility of payment by smartphone. So, (AI) technology will impact human's daily life consumption behaviors in the future. For another example, transportation tools, such as boats and ferries and private vehicles will use sensors and navigating without human input. Taxi and truck drivers will become obsolete, the stock store applies to stock managers and postal carriers of the delivery is distributed by (AI) machine delivery method.

What is the relationship between (AI) and (CRM)?

- Can (AI) technology impact on customer relationship management (CRM) ?

Nowadays , (AI) is a technology almost as old as the computer industry itself,

it is similar with the advent of personal assistants function to businesses and personal promotion channel, such as ( Amazon's Alexa, Apple's Siri, Google's Assistant) image recognition ( face book), personalized recommendations ( Netflix , Amazon). Those innovations have been driven by a increase in processing power, lower cost hardware, and the exploding creation and availability of data. It seems, (AI) technology can impact global customer service management method.
How to forecast economic impact modeling to (AI) will affect global economy? Can human forecast business revenue growth and job creation ( or destruction) based on (AI) applied to customer relationship management (CRM) activities? In addition to the economic impact on (AI) or (CRM) which can include an estimate of the economic impact attributable to sales forces customer base. What can economic benefits be brought to (CRM) from (AI) technology?
Artificial intelligence(AI) comprises a set of technologies that use natural language processing, machine learning, knowledge graphs, and other tools to answer questions, discover insights and provide recommendations. Computer systems can use (AI) hypothesize and formulate possible answers based on available evidence can be trained through the ingestion of vast amounts of content, and automatically adapt and learn from (AI) self mistakes and failures.
So, any business organizations (customer service departments) can provide efficient and effective customer relationship management of excellent customer service quality if which applied (AI) technology system. The different type of (AI) systems include: (AI) system platforms, machine learning (AI) based data preparation and enrichment tools, machine vision/ image recognition, voice speech recognition, text analysis and natural language processing, bots , e.g. face book website and virtual digital assistance solutions, social media pattern analysis , sentiment analysis, advanced numerical analysis (e.g. IOT streaming , machine logs), supporting technologies, knowledge base dialog management, Q&A processing etc. different (AI) technology system customer relationship management (CRM) tools.
(AI) (CRM) of activity can include these categories, such as: corporate marketing, marketing operation, field marketing, customer support, digital commerce, customer analytics, customer influenced product or service design, product or service pricing, finance information, presentation, customer billing, inventory , logistics and fulfilment support, partner

management etc. different CRM tools.
(AI) technology of CRM has been carrying on plan different stages to achieve CRM personal assistant tool for businesses. The stages are such as, in the beginning stage of (AI) projects in place, implement now, pilot phase next year in the final stage of (AI) customer relationship management tools are foreseeable future. So, this CRM technology has been improved to plan in different stages every year to prepare to achieve full capacity of CRM service quality for businesses to use in the future.
Hence, how to develop an estimate prediction of the economic impact (AI) technologies could have CRM activities, which depends on gathering macroeconomic information on business revenue and the basic marketing of business revenue and the basic markup of business expenses by major functions ( customer support, marketing and sales , production etc.)
An economic impact model that can gather data together and forecast the results how (AI) artificial intelligence technology brings (CRM) customer relationship management benefits to businesses, e.g. surveys investigation includes IT spending by sample countries, GDP and population estimates and forecasts, revenue per employee and ratios of IT spend to GDP. Surveys ( questionnaire questions) of forecast results are influenced by (AI) impact can include: results are projected from surveys and rely on estimates are made by respondents on the expected financial improvements in categories of (AI) –assisted customer relationship management activities. The forecast assumes that these estimates are correct; financial estimates are based on estimates of “first year” improvement from full (AI) implementation; forecasts are from planning to implement any artificial intelligence of customer relationship management (CRM) projects, the improvement forecast is of categories of activity , e.g. corporate marketing , digital commerce, and customer analytics. They are not estimates of ROI for the (AI) software. They rely on conservative estimates to which each of these entities might affect company revenue, expenses or productivity. They also rely on estimates of the penetration of software in customer relationship management activities . Net new jobs created are based on the ratio of new revenue to jobs required to support that revenue . They can assume that 50% of the net new revenue will support increases in labor and the rest will go for capital and other operating expenses that may replace jobs lost to automation.
In the future, some of the ways in micro economic benefits to any organizations. (AI) technology is expected to impact CRM activities include:

Spending up sales cycles, improving lead generation and qualification solving customer support problems faster ( raising service quality), helping companies improve brand campaigns and recognition, lowering costs of support calls when increasing resolution rates, lowering the cost of recruiting employees and partners, increasing revenue from optimized product marketing, optimizing price, distribution logistics and preventing loss through fraud detection. So, micro economic benefits view point, it seems that (AI) CRM technology can raise any companies economic benefits for care term.

Artificial intelligence enables machines or the in-build software to behave like human beings which allows these decisions and act. The advent of (AI) is leading , talking, making decisions and act. The advent of (AI) is leading to new technologies advances and transforming the economic and employment opportunities for humans in a positive way. (AI) related technologies can facilitate our live. For example, industrial robotics, robotic medical assistants, smart games, financial forecasting software, big data analysis, algorithms in health and bioinformatics, pilotless cargo places, drone ambulances and general purpose and workplace robots and others. (Disruptors technologies: Advances that will transform life, business and the global economy).

Artificial intelligence also known as computational intelligence is defined as " the human –like intelligence exhibited by machines or software. It is theorized that intelligence of humans can be described and intelligence machines or software can simulate it. These machines software can be reasonable , learn, perceive and process information, like human mind and thus facilitate human life. They can think and act for us. So, artificial intelligence is an interdisciplinary field of study including computer science, neuroscience, psychology, linguistics and philosophy.

However, (AI) research and developments have economically impacted many industries, such as robotics, telecommunications, computer applications , health, finance, heavy manufacturing, transportation, aviation, e-service and e-commerce, military , music and movie, toys and games entertainment etc. industries.

In fact, many ideas, systems and technologies have been developing in the world of (AI) technology. However, which are net called or considered (AI) products, rather which are mentioned with their specific names, such as smart graphics, machine learning, e-commerce etc. ( i.e. this is called (AI) effect).

● How can (AI) technology influence digital economy?
Nowadays, (AI) related industrial applications will replace most human power in fields, including call centers, customer services and air cargo transportation. (AI) technologies also help weather forecasting based on repeated rainfall pattern ( data) recognition, through robotics ( i.e. floor cleaning, moving lawns etc.) transporting people and products with unmanned vehicles, sending space unmanned smart shuttles, developing robotic arms, predicting market values in stock exchanges by internet, making homes safer, helping elderly and disabled using robotic servants etc.

Among the (AI) related technologies , there are a few that significance for the impact on society and especially on digital economy . (AI) is particularly influential in machine learning. Such as robotics, transportation, finance, health and bioinformatics, e-commerce , e-games, big online data gathering and internet-of-things. For example, machine e-learning is based in bioinformatics and robots that can learn new skills for better caregiving in healthcare. What is machine e-learning? Machines can e-learn from e-data gathering, coming up generalizations and making decisions to act in certain ways from internet.

There are important applications , such as e-machine perception, electronic online natural language learning processing, online search engines, online bioinformatics, online brain –computer interface, online game playing, online robot locomotion, online advertising, online computations finances, online health monitoring, online DNA classification and decision making, online in chemistry –cheminformatics . So, online machine learning can positively impact productivity and it can enhance information and analytical system from (AI) online channel.

What is robotics? Robotics is one of the most strongly influenced fields in (AI). For example, heavy manufacturing industries, robots and used and man power is replaced for effectiveness, precision, and accuracy, especially in respective or dangerous tasks, including welding, assembling , picking and placing .

So, robots can acquire new skills or adapt the changing dynamic environment. Also, artificial intelligence can be applied in developing transportation. For example, automated vehicles, driver assistance systems , safety systems, collision avoidance systems and public transportation. Moreover, (AI) technology has proven to produce some of the best tools to predict stock market fluctuations from internet data gathering method. It's predictions are based on ever-evolving predictions algorithms and systems learn new models and make connections between historical data and new data to measure stock market trading more accurate from internet data gathering channel.

In health field, especially in health data processing , analysis, decision making support and medical diagnosis. So, online data can show which patients will need what treatment and what alternative drugs could be used more accurate from (AI) online data gathering method. Bioinformatics is an interdisciplinary field combining statistics, (AI) online technology can help in discovering data patterns and modeling through the application of machine learning, artificial neural networks and genetic algorithms. For example, further (AI) technology development of human genome project of online data sequences.

Online shopping can be facilitated by virtual assistants developed through (AI) technology and these assistants can offer the best advice. (AI) online purchase coming after every product image recommendations and personalization bring important revenue to shopping online sites, like Amazon . Smart computer graphics and games, artificial intelligence is useful in smarter computer, graphics, scene modeling , scene rendering processes in order to create, for example, effective human –robot interactions , online machine learning, online strategic games techniques etc. online computer related (AI) software.

So, online big data analysis and big data does have a critical need in the world of online intelligence machines and software in our future. In other words, (AI) offers online technology to enable online big data analysis to provide industrial organizations with valuable information for effective decision making in short time. For example, what IBM's Watson achieved: this machine used 200 million of structured and unstructured content with a special technology of hypothesis generation, massive evidence gathering, analysis and scoring from internet channel.

Finally, (AI) online technology another related internet invention ( internet of things) (IOT) is the network of machines or objects connected through internet. These connected objects can sense their internal and external environment, communicate with each other, can send critical data and finally can make decisions to act or correct their environment from (AI) online technology. For example, factories can monitor and automatically change production processes, hospitals can monitor and regulate the health

conditions of their patients , schools can collect data from facilities and cars can send data to car makers from (AI) online technology.
Partner predicts that (IOT) market will create about trillion amount value by 2020 year. Although machines collect big data from their environment, whether which gain an insight or learn from these online data largely depends on the (AI) online machine learning principals and (AI) online technology. In 2013, Mckinsey estimated that disruptive technologies closely related with potential economic impact in 2025 year between $7.1 to $13.1 trillion amount ( automation of knowledge work, advanced robotics, autonomous or near-autonomous vehicles).

What is the relationship between
(AI) and global digital economy development ?

● Could work activities in China be automated
making in the nation with the world's largest automation potential?
Can (AI) technology influence China economy? Could China workers be affected and jobs made up of routine work activities and predictable? Will programmable tasks be particularly impact to China employment market ? When impact on labor market is likely to be gradual at the aggregate level, it can be sudden and dramatic at the level of specific work activities, rending some job obsolete fairly. Overall (AI) technology will raise digital skills when reducing demand for medium incomer inequality for China workers. It seems (AI) technology's effect on productivity could be crucial to China's future economic growth as the population ages are increasing.
In China, some biggest technological companies driving significant investments in research and development. Moreover, China is one of the leading global (AI) technology development county. However, China will need to focus on building its innovation capacity. For example, United States and United Kingdom are currently producing more influential (AI) technological research. However, if China planed to achieve (AI) technology success, it's traditional industries will need to develop technical know-how –to and overcoming implementation costs prepare to develop (AI) . When (AI) technology is introduced into China society, China government needs to raise concerning ethical, legal, technological security etc. business questions. Also, surrounding issues include privacy, discrimination, legal liability and regulation. It aims to encourage overseas investors to choose to invest (AI) technological industry to raise GDP growth and manufacturing industries income growth for long term in China.

If China encouraged overseas (AI) technology investment in its country. It is possible to influence China employment market to be changed. Because (AI) technology will impact to influence China people daily life. Due to (AI) technology is introduced to China society, many rich people will prefer to spend to buy any high (AI) technological products for entertainment or learning or machine man driving etc. daily necessity activities. Then it will raise GDP growth and will raise (AI) manufacturers or related-(AI) technological manufacturers profit. It is beneficial to China because it can become one high knowledgeable and (AI) technological economical society. But it will bring bad influences to raise unemployment chance for the low skillful labor. In labor economy aspect influence , how (AI) technology can influence China low skillful labor unemployment ratio raising. The raising low skill labor unemployment reason is because China low skillful human labors are argued or are replaced by (AI) technology creating new challenges to introduce to influence China society of simply human manufacturing job nature to be changed to be high (AI) technology manufacturing job nature in any China factories. Moreover, when (AI) technology introduction to China, it will cause other related social challenges in China. The varied (AI) related challenges, including the difficulty of creating safe and reliable hardware for sensing and affecting ( transportation and education), the challenges of gaining public trust, a low resource comities and public safety and security, the challenges of overcoming fears or marginalizing humans in China employment and workplace and the risk of diminishing interpersonal trust because the low skillful labors won't believe any China employers will give chance to employ them , due to (AI) technology will replace their skills and man manufacturing of productivity is much less to compare to (AI) technology manufacturing method.

- How does ( AI) technology influence
the future of employment change?

Are future nature of jobs changed to computerization from (AI) technology? Where are the probability of computing occupations from (AI) technology influence? What is expected impacts of future computing on labor market from (AI) technology influence? John Maynard Keynes's frequently cited prediction of widespread technological unemployment " du to our discovery of means of economic the use of labor outrunning the pace of which we can find new used of labor" ( Keynes, 1933, p.3).

In the future, (AI) technology will impact some nature of occupations to change computing. This chance will also influence some countries' economic change. For example, some factory human labors hand routine manufacturing tasks will be changed to computerization of routine manufacturing tasks by (AI) technological machine men hand manufacturing method. it will cause a structured shift in the labor market, with workers reallocating their labor supply from middle-income manufacturing to low-income service occupations.

Arguably, this is because the manual tasks of service occupations are less computerization, as who require a higher degree of flexibility and physical adaptability. So, (AI) technology will influence the human hand labor skillful occupation nature of task cheaper , such as vehicle manufacturing , ship manufacturing, computer manufacturing, steel manufacturing, television, radio etc. home electronic products of heavy machine industry change. Due to (AI) technology machine man will be proper to be used to manufacturing these electronic products when the (AI) technology innovation can develop to the mature stage. Then, any countries manufacturers will choose to use (AI) technology machine man, instead of human hand production.

Supposing the future prices of computing are fallen, seriously, problem solving skills are becoming relatively productive, explaining the substantial employment growth in manufacturing occupations, involving cognitive tasks where skilled labor has a comparative advantage, as well as the increase education needs for (AI) technology computing of machine man subject study.

Prediction of education needs for (AI) technology student numbers will increase, due to manufacturing industry needs many (AI) technology students in future employment market. Another (AI) technology influence if the future (AI) technological innovation, e.g. machine man manufacturing or machine man service industries will both increase demand, then with more sophistic software technologies will be disrupted labor markets by marketing workers redundant.

For publishing industry, what is striking about the case in paper book publishing industry will be unpopular? Due to the electronic book publishing industry will be popular, e.g. Amazon publish . (AI) technology can influence paper book manufacturing method which is replaced by machine man electronic book manufacturing method as well as it will cause the computerization is no longer confined to routine manufacturing tasks. Due to (AI) machine man manufacturing technology will be proper to be used to manufacture any products in short time efficiently and effectively , e.g. electronic book products. In the future, if it is fact to occur this case, such as ( AI) technological machine man manufacturing method will be adopted ( applied) to manufacture electronic books or any products in possible. (AI) technology will cause many manufacturing workers are unemployed. It is beneficial to employers, who can reduce to spend much wages expenditure to employ manufacturing workers, but it will cause many manufacturing workers loss jobs and reduce income to support whose families lives. It will cause social challenges, e.g. increasing stealing crimes if the manufacturing workers had not other skills to find other jobs to do easily. So, manufacturers need to concern over technological unemployment which will be hardly future phenomenon if who decided to dismiss all manufacturing workers, due to (AI) technology machine men replace to them.

If ( AI) technology can be innovated to produce any kinds of machine man to serve any service or manufacturing industries successfully. Then, it will bring these questions: Can future that workers be influenced to be automation employment and productivity by (AI) technology influence? Does it impact to influence the (AI) technology countries' productivity and growth and natural resources development and labor markets and evolution of global financial markets and economic impact of technology and innovation and urbanization etc. issues? How will automation transform the workplace? What will be the implication for employment? What is likely to be its impact both on productivity in the global economy and on employment?

In fact, automatic of activities can enable businesses to improve performance by reducing errors chance and improving quality and speed, and same cases achieving outcomes that go beyond human capabilities. Some economists indicate (AI) technology would give a needed boost to economic growth and prosperity have of the working age population in many countries. Based on the scenario modeling, they estimate automation could raise productivity growth globally by 0.8 to 1.4 % annually. They also indicated that almost half the activities people are almost $1.6 trillion in wages to do in the global economy have the potential to be automated adapting current demonstrates technology, according to their analysis of more than 2,000 work activities across 800 occupations. When less than 5% of all occupations can be automated entirely using demonstrated technology, about 60% of all occupations have at least 30% of worker made activities, that would be automated. More occupation will change to be automated. They also indicated for business performance benefits of automation are relatively clear, but the issues are more complicated by policy making to attract foreign investors. Beyond technical feasibility, the cost of technology, competition labor will include skills and supply and demand dynamics, performance benefits and beyond labor cost savings and social and regulatory acceptance will affect the automation. Their predictions suggest that half of today work activities could be automated by 2055 year, but this could happen 10 to 20 years earlier or latter depending on the various factors in addition to their wider economic condition.

Some scientists suggest (AI) technology is finally starting to deliver real-life business benefits. Computer power is growing significantly , algorithms are becoming more sophisticated and perhaps most important of all, the world is generating vast quantities of the fuel that powers (AI) technology data billions of gigabytes of it every day. Also, online firms are digital natives, such as Google online search service company is investing on (AI) technology. For new though most of the news if coming from the suppliers of (AI) technologies. And many new users are only in the experimental phase. Few products are on the market or are likely to arrive these soon to drive immediate and widespread adoption. As a result, analysts believe (AI) technology's potential will give true economic benefit in the future. (AI) industry will introduce to suppliers and users to raise economic potential of (AI) technology.

In the future, (AI) technology systems can solve business problems. Some scientists categorized those into five technology systems that are key areas of (AI) technology development: robotics and autonomous vehicles, computer vision language virtual agents and machine learning , which is based on algorithms that learn from data without replying on rules-based programming in order to draw conclusions or direct an action.

Such as computer vision and language includes natural language processing, analytics, speech recognition technology, some are about learning from information, such as about machine learning and others are related to acting on information, such as robotics, autonomous vehicles and virtual agents, which are computer programs that can converse with humans. Machine learning and a subfield called deep learning are artificial intelligence applications.

- Can artificial intelligence impact
global economy growth?

Artificial intelligence (AI) is a term first defined in 1956 year. It is a branch of computer science that aims to create intelligent machines that work and react like humans. In contrast today, 60 years later, (AI) is characterized by a number of applications, including computers playing games against humans and understanding human languages, virtual personal assistants, and robotics which involve computers seeing , hearing and reacting to sensory stimuli. In the future, technologists predict for (AI) technology ranging from (AI) being used as a tool to aid relatively simple processes for robots with human like mental capabilities, who expect (AI) technology can emulate human performance by learning, coming to mind its own conclusions, understanding complex content, engaging in dialog with people, enhancing human cognitive performance or replacing humans in executing both routine and non-routine tasks. In existing industry, (AI) technology is used , such as targeted advertising and virtual used personal assistant as well as the (AI) technology that my exist in the future, such as robots with human vehicle processing capabilities.

The range of (AI) technology's progress in the future will determine the economic impact future of (AI) technology on the global economy with more limited advances and applications ( i.e. weak (AI) only) corresponding to more limited economic impacts and more substantial progress, i.e. strong (AI) technology is corresponding to more significant economic impact.

(AI) technology learning that automates analytical model, including predicting cause-and-effect relationship from biological data, identifying new drugs, self-driving cars and protecting against fraud etc. functions. Also (AI) learning can improve natural language processing that allows computers to continue to better analyze, understand and generate language to interface with human using the natural human language, virtual personal assistant, helps users by providing scheduling appointment, reminds organizing personal finance and finding providers of various services, machine vision allows (AI) machine man to identify object, scenes and activities in detect pedestrians and bicyclists.

We expect the economic effects of (AI) technology to include both direct GDP growth from sectors that develop or manufacture (AI) technology and indirect GDP growth through increased productivity in existing sectors that employ some from of (AI) technology. If (AI) producing sectors could grow, then it could lead to increase revenues and employment of (AI) technological professionals within these existing firms as well as the potential creation of entirely new economic activities to any countries' societies productivity improvement in existing sectors could be realized through faster and move efficient processes and decision making as well as increased (AI) technological knowledge and access to information available in societies easily.

In the future, if (AI) technology is an increasingly critical component of more products, it will become an integral part of necessary products of many people's lives. The extent of (AI)'s economy effort is also likely to vary from region to region, thought variation may be more dependent on the predominate economic activity of a region and the (AI) ability can influence economic activity, rather then the economic or developmental status of the regions. (AI) technology can move accessibility and can use source development to do international business between one country and another country.

So (AI) technology has the potential to give benefits to different income chooses and to bring significant gains to both developed and developing countries. For agricultural technology, (AI) has the potential to optimize food production around the world by analyzing agricultural regions and identifying what is necessary to improve crop yield. In total, (AI) technology gives greater economic impact to any countries agricultural regions if which implemented (AI) technology to grow crop , fruit etc. food production in the farms.

Investment in (AI) technology is such as capital investment to any countries' public or private enterprises. So, it will have large economic impact to the future . If the (AI) technology is reasonable invested to the different needs aspect by the public or private enterprises in the country. Then, it will have good economic impact to the country in the future. However, when (AI) technology is likely to affect both the productivity and employment components of economic growth in many sectors. Significant public debate has focused on projections of (AI)'s effect on the labor force. However, for instance, some researchers have argued that the rise of (AI) technology and automation will led to significant unemployment as capital is substituted for the low skillful labor. So, they point to the concern that the increasing sophistication of (AI) technology may balance skilled and semi-skilled workers and the reduce the size of the middle class. However, this is not a new argument, due to (AI) technology negatively affecting the labor force and leading to mass unemployment. Because the (AI) technology is the substitution of machinery for human labor. Although, employment in certain industries, has been reduced in the past due to technological advancement. For long term, the labor market has adapted to the introduction of new technology, giving rise to new jobs in new areas. (AI)

technology may also be accomplished without a reduction to total employment in the long-term to some Asia countries, such as Hong Kong and Japan. Because Hong Kong and Japan many low skilled labor, e.g. security, cleaner who complaint that employers need them to work long time hours. ( abnormal working hours) e.g. one day 12 to 15 working hour per day. Hence, if (AI) machine means invention technology success. Security or cleaning job can be worked from (AI) machine man in some hours every day in order to reduce the long time working hours cleaners or security workers, e.g. one (AI) machine man works 4 hours for cleaning or security job, one day as well as another cleaner or security labor only needs to work 8 hours one day. So total security or cleaning employers can employ 12 hours machine cleaners or security workers and human cleaners or security workers in one day. For long term benefit, Hong Kong or Japan every security or cleaning worker does not need to work 12 hours minimum working hours one day. They won't feel tried and bore and without private with whose families, so who will accept to do these cleaning or security jobs, even they can raise work efficient and performance when who feel happy and health.
So, (AI) technology of machine man invention can raise low skillful labor efficiency and it can help them to avoid abnormal working hours demand in some busy work life countries, such as Hong Kong and Japan. Before, one Japan female labor feel unhappy to work, due to who often needs to work abnormal working hours for her employer and who has less sleeping and without any private time to enjoy her life with her families every day. So this abnormal working hours factor causes her to do commit suicide behavior, then she is die unlucky. So (AI) technology of machine man invention ought avoid abnormal working hours demand for employer in any countries in the future.
The most important occurrence to any employers, some researchers had attempted to do one experiment to find that private research and development , venture capital and public research and development investment all have strong net effect or economic growth with venture capital funding further having the strongest such effect from (AI) technology. The researchers hypothesize the venture capital investment contributes to economic growth through (AI) technology innovation and by the capacity of an economy to use existing (AI) technology knowledge to increase productivity. They predict the impacts of venture capital, business-research and development and public research and development can raise multi factor productivity from (AI) technology introduction.

Can (AI) technology influence the economic development to developing countries? The developing regions of the world contain most of natural resources. If one day, (AI) technology has invent one kind of machine man which can assist any gas or oil workers to seek any new oil/gas natural resource locations easily. I believe that (AI) technology can help these natural resource exploitation countries will gain economic benefit more easily. So, (AI) driven technology can be used to change to create any new opportunities to address poor management or resources and improve human well being, such as Africa Latin America and India can use (AI) technology machine man to seek any oil/gas natural resource countries exploitation activities to attempt to gain much economic benefits.

● Why will (AI) technology grow economic development ?

Nowadays, increases in capital and labor are no longer driving the levels of economic growth, such as (AI) technology. The ability of increase in capital investment and in labor of traditional drivers of production, have no longer to be enjoyed in most developed economies ,e.g. developed country, US, UK . However, artificial intelligence has the potential to overcome the physical limitation of capital and labor to avoid missing out on this opportunity. So, policy makers and business leaders must prepare for and work toward a future with artificial intelligence. They must do with the idea that (AI) is another simply method to enhance productivity method . Rather they must see (AI) as the tool that can transform thinking about how growth is created.

Economists have always thought of new technologies are as driving growth their ability to enhancing. It can replace labor and capital factor of production. So, it brings this question: What is the factor of production (AI) technology characteristics. They key factor is to see (AI) technology as a capital-labor .

(AI) can replicate labor activities at much greater scale and speed, and to even perform some tasks began the capabilities of human. For example, by using virtual assistants , 1000 legal documents can be reviewed in a matter of days instead of taking three people six moths to complete. Some (AI) technology may be one kind of factor of production in the future. For another example, people will work in workplace digitalization environment. So, in the future, working environment and information management are automated. Such as Konica camera sale company will use workplace digitalization. So , (AI) technology can provide workplace digitalization in order to raise productivity efficiency. (AI) technology will be one kind of production which is replaced by workplace digitalization and it will grow any organization productivity efficiently. Then, (AI) technology will assist overall social economy growth , due to productivity is raised and products can be produced in short time to prepare to sell in consumption market. So, time will be shortened to increase GDP growth fast for the development of (AI) technology countries.

● How can (AI) technology impact to global economic and social and psychological changes?

What will be the development of (AI) technology and predictions concerning the future evolution? The computers and robots will develop conscious, intelligent and minds into humans, enhancing psychological and behavioral abilities and allowing for direct communication with (AI) minds. (AI) technology will be impacted human life by (AI) technology information communicative and environmental influence. A " world brain" and " world mind", this psychological system will be enhanced and enriched the capacities of both individual and collective cognition by (AI) technology of service industries.

(AI) technology with influence these human needs of service industries changes, such as , biological science, finance, entertainment, business, biological science, transportation, communication military etc. The personal computer evolution, the internet and the world wide web which exploded on the scene, linking business, homes, schools, social organizations which were a completely unpredicted phenomenon to influence human life. Kurzweil (1999) predicts that by 2029 year, most human communication will be with machines. According to Person, by 2100 year, there will be human machine convergence.

How can (AI) technology influence environmental protection to make benefits to farming economic growth? (AI) technology can be applied to predict how to solve environmental pollution challenge to avoid to damage any crop or vegetable or rice or fruit etc. food growth. Because environmental experts can gather global environmental pollution data from an environmental database to build a perform a systematic analysis from (AI) technology. The first step is this broad analysis can include understanding, statistical and data gathering techniques to obtain the relevant data, the correlation among the variables involved, and a list of possible models. The next step is to select a set of methods and models that cover all kinds of knowledge and functionalities needed for the decision making process. Once the models are selected, they must be fully implemented by means of machine learning , data mining, statistical or numerical technique. After that, those models must be integrated to build the whole EDSS. The EDSS must be tested to check its performance, accuracy, usefulness and reliability, both from the user's and (AI) technology/computer scientist's point of view. If these is any wrong feature in any development stage, such as model's integration, models' implementation, selection of models, database, problem analysis etc. the developers must come back in the update th required components. When the evaluation phase is all right, the EDSS is ready to be applied to the environment. The great contribution of artificial intelligence to EDSS the integration of several methods complementing the classical statistical models/simulation , statistical analysis, linear models, etc. and numerical models ( control algorithms, optimization techniques etc.) .

This cooperation makes the resulting systems more reliable and powerful in coping with real world environment systems. Date interpretation has been a principal area of research in (AI) technology since the very beginning. The most demanding problem in the environmental assessment context. Knowledge representation permits the definition of the different types of data that the existing methods adapt to the process. There is also a lot of work to clean, repair and transform the huge available quantities of raw data. Apart from this, the availability of meta-information or background knowledge is required to guide the process. Data mining is multi-disciplinary: It covers expert systems, data based technology, statistics, data visualization and unsupervised machine learning. These techniques operate at the level of data and background information, where numerous and often incompatible new commensurate pieces of information from disparate sources have to be brought together ( K, Fedra, 1994).

So, it seems that in the future, (AI) technology with the increasing maturity in particular those related to knowledge and engineering, new dimensions can be assisted to users in environmental decision making are available. For example, many environmental systems are characterized both by incomplete models and by limited data. Hence, in the future, (AI) technology will be applied to predict climate change to reduce crop or fruit etc. food agriculture challenge by climate change bad influence.

- Will (AI) technology influence digital economy change to manufacturing industry ?

To understand how the manufacturing business must adapt to prosper in the technology, we need to understand how (AI) technology will change us to shape our daily habits to satisfy our expectation of products to how we shop and even the immediate of the entire process. For example, taxi services are in the crosshairs as on demand transportation services like, available of the touch of a smart phone button expand. In fact, Yellow lab, US country , san Francisco city's largest taxi company is filing for bankruptcy as the industry starts to change faster than almost anyone expected. However, at this point, its more than an app that is changing, some our taxi passengers renting taxi transportation to catch consumption behavior.

(AI) technology will influence digital economy for taxi passenger's individual customer experience, offering a growing renting taxi to catch of service and feedback opportunities when any one taxi passenger who chooses to use mobile phone app online tool to prepaid to rent any taxi more easily.

Also in the long term, (AI) technology can influence vehicles drive themselves of behavior. Already, companies like Google and GM are working on projects to bring fleets of autonomous vehicles to cities at the path of a button.

Moreover, this on-demand service model is beginning to appear across a much broader range of markets. For example , Amazon company is investing in its own fleet of trucks, planes and even drone at the same time as it pushes for same-day delivery of products. As some point, vehicles will be autonomous too. So, it seems that (AI) technique will influence any transportations choose to use digital autonomous driving technology in the future . For Amazon company case, it is not stopping of logistics. It is also aiming to automatically manage the supply of consumer home products with its recently launched Amazon replenishment service, Dash. Dash is a digital service that enables that connected derive to automatically order physical products from Amazon when supplies are running low. So, it seems (AI) technology will be applied to logistic function by digital technology method introduction in the future.

Hence autonomous vehicles will optimize industry supply chains and logistics operations through increased efficiency and flexibility. In fact, fully automated and lean supply chains will keep reduce load sizes and inventory by leveraging smart distribution technologies and smaller autonomous vehicles by machine man assistance. If Amazon continues to grow market share for online sales by reducing effort required by the consumer to place an order, when also contributing the almost immediate delivery of products to the doorstep. So, it will further fuel the trend toward on-demand derive. As Amazon company fuels the on-demand economy, consumers will expect immediacy in more parts of the digital economy. On top of speed, consumers increasing expect more personalization options.

So, (AI) technology will influence digital manufacturing, such as Amazon publishing to monitor every aspect of every process in real -time and communicating to self-optimized deep learning robotics, new methods of high volume and high customization will become possible. Then, as products merge into product platforms and even services, manufacturers have the opportunity to provide components and platforms used by smaller players. So, (AI) technology will influence manufacturing industry to choose automated SMI lines, robots installed, automation engineers.

Another future (AI) technology development can be applied to space science aspect, such as Automation engineering space in manufacturing process to achieve digital manufacturing benefits to any businesses in the future. Such as reducing cost, shortening manufacturing time, raising efficiency, shortening delivery products to client individual time. How can artificial intelligence give the need and advanced fast and evaluation methods benefits for space exploration? When US NASA ( space exploration organization) achieves any space exploration missions, it will answer this question:

When is it useful to have a machine use (AI) technology to achieve a decision? After all, after millions of years of space exploration and rough 10,000 years of civilization, humans are usually quite good at making decisions in complex uncertain environments. Through, Johns Hoplains University's Applied Physical Lab. Research in (AI) technology enabled systems, which has identified three general use cases for (AI) technology to explore space mission:

First, for some tasks (AI) technology is more cost effectiveness than human. Second, (AI) technology is better suited than humans at solving some, but not all problems. Third, (AI) technology allows NASA organization's space exploration mission to develop machines that ate capable of responding faster than when a human is in the decision loop ( D. Scheidt, 2012, A. Castano et. al. 2008).

So, the use of (AI) technology to enable science by observing the pace of rapidly evolving phenomena was demonstrated. It is more effectively coordinating and (AI) technology utilizing to earn economic benefits to use for space exploration mission.

However, (AI) technology also have current risk for space exploration. Today (AI) technology is immature and requires further development to reach its potential. For instance, the (AI) technology algorithms that detected the dust derive could not have identified whether the Martain weather represented a threat to the cover. Also it can not yet use instrument input to determine what, where and how to autonomously make the next space science measurement. An equally important factor limiting (AI)'s deployment is that lacks the methodology and technology to effectively test (AI) technology. So, the challenge will testing (AI) enabled system is how (AI) performance can be measured. It would be NASA organization's difficulty to find (AI) technology to develop to carry on researching any space exploration missions in the future. However, (AI) technology will be a good economic benefit choice for space exploration mission in the future.

- What is artificial intelligence potential benefits and ethical considerations?

The ability of (AI) technology systems to transform vast amounts of complex information into insight has the potential to help solve manufacturing or service challenges for human needs. However, to reap the societal benefits of (AI) systems, humans will need to trust then and make sure that which follow the same ethical principles, moral values, professional codes and social norms that we humans would follow in the same scenario, research and educational efforts as well as carefully designed regulation in order to achieve the most effort of economic benefits goals. For example, international business machines corporation (IBM) is actively engaged both competitors , in global discussions about how to make (AI) ethical and as beneficial as possible for people as social economic benefits.

(AI) is usually defined as the " capability of a computer program to perform tasks or reasoning processes " that human usually associate to intelligence in a human being. Often, it has to do with the ability to make a good decision, even when there is uncertainty, too much information to handle. As an example, play chess or complex card games of entertainment activities is believed to need some form of intelligence in a human being, as well as choosing the best medical facilities in a difficult medical case, or creating something new, such as mathematical theorem or even some form of act, or even driving automatic machine man ( self driving vehicle) replacing human driving in the middle of a crowded city.

(AI) needs depends on what we consider being intelligence in the behavior of a human being act a certain point in time. If human belief about human intelligence changes and we don't believe any longer that a certain task requires intelligence, then a computer program performing that task is no longer part of (AI), it becomes just another boring computer program. So, it means that (AI) technology will replace some old computer programs, if human can invent new generation of (AI) software for any functions or activities to satisfy human needs.

As IBM, it argues intelligence. This means that we aim to build systems that enhance and scale human expertise and skills rather than replacing them. We therefore focus on practical applications of (AI) capabilities that assist people in performing well-defined tasks of needs by exploiting and wide range of (AI)-based services. We also use the term " cognitive computing" it is mean a comprehensive net of capabilities based on technology. It comprises the fields of machine learning, reasoning and decision technologies, language, speech and vision recognition and processing technologies, high performance and high efficient functions for any industries or individual consumers needs. For example, robotics, which are usually very good at doing what which are supposed to in any environment, much have public shopping center, factory etc. places which need simply services from the robot ( machine man), such as cleans the floor of our houses to the robot that can work together with humans in production chains, passing through the warehouse, robots can take care of the tasks of an entire warehouse and the companion robots like Nao, Pepper, Aibo and Giraff, who can entertain use, talk to use and help elderly people to stay connected to their friends, relatives and doctors.

Google company is building automatic machine ( self-driving cars ) and has acquired more than 10 robotics companies. Facebook had opened whole new research facility only on (AI) research. Apply computer has developed Siri. Microsoft computer company has built a similar personalized assistant. Google has Deep mind, a UK company whose long term aim is to build general (AI) and has already great potential to win game to the world champion and IBM is investing a huge amount of resources in applying its Watson cognitive computing system to the medical domains to finance and to personalized education. In Europe, IBM is establishing new centers in Munich and Milan focused in the application of cognitive computer

capabilities to the internet of things and healthcare respectively.
For example, automatic machine man ( self-driving cars) are all about (AI), which used to be able to see what happens in the street ( signals ,lanes, other cars, pedestrians, traffic lights, which need to able predict what other cars and pedestrians will do, and who need to be able to cope with unforeseen situations. Since, most car accidents are due to human fault, it is estimated that the adoption of self-driving cars will save about half of the lives that are usually last in car accidents.
IBM Watson company has to understand spoken language, make sense of massive amount to text , respond correctly to questions in many categories, as well as assess its own confidence in responding to such questions. In the future, (AI) technology can own question/answering capabilities that would be very useful, for example, in assisting a doctor when trying to some to the correct diagnosis for a patient and to propose the best therapy .
Intelligent machines can also rely on huge amounts of data to be used to learn how to make better decisions. This data comes from all of us over the years Facebook users have uploaded more than 250 billion pictures and every day who upload about 350 million more. Every second, we submit 40,000 google search queries. So, (AI) technology will be connected through the web from appliances to traffic lights from cars to watches. Other tasks that are very easy for humans are physical and manipulation tasks, such as walking , running, picking up an object to make its shape and location, restricted environment. But (AI) machine man technology still not able to have the general physical and manipulation capabilities even of a 6 year old.

So, it brings this question: Why do (AI) scientists need to concern ethics? Because (AI) technology is complex, information into insight has the potential to reveal long held secrets and help solve some of the world's most difficult problems. (AI) systems can potentially be used to help discover insights to treat disease, predict the whether, and manage the global economy. So, ethic issues is important to and (AI) scientists . If any one new (AI) technology research investigation could success, it will be a secret to and the (AI) scientists can not permit to their loyalty to any competitors to damage the fair (AI) technology products trading market. The country ( countries) (AI) technology scientists need to concern ethic issues, who need to keep secrets for their countries economic or/and social benefits. This is moral issues to any countries/country loyalty is whose countries intangible assets. They can not sell (AI) loyalty to any their countries to assist whose economic benefits immorally.

- How can (AI) technology influence to global health care economy development?

According to (AI) lecturer analysis, when combined key clinical health (AI) application can potentially create $150 billion in annual savings for the US healthcare economy by 2026 year. (AI) technology is re-winning modern conception of healthcare delivery. It enables machines to sense, comprehend, act and learn. So which can perform administrative and clinical healthcare functions (Accenture, 2017).

It will help health care service organizations to reduce health care cost, will improve and raise service quality and access. So, (AI) health market size will be predicted growth. (AI) applications in health care include robot-assisted surgery, virtual nursing assistant, administrative workflow assistant, fraud detection, error reduction connected machines, clinical trial participant identifier, preliminary diagnosis, automated image diagnosis and cybersecurity.

What kind of benefits (AI) technology can contribute to healthcare service? (AI) technology can deliver what many health care organizations need, such as financial and operational of labor costs, digital expectations from patient consumers how to use (AI) technology to solve interoperability challenges in any healthcare organizations. Also (AI) technology can be applied to wellness an d lifestyle management, diagnostics, delivers financially but also way of organizational and workflow improvement. So, (AI) technology will be continue to become most prevalent and adoption to healthcare organizations , which must need to enhance structure to be position to take full advantages of new (AI) technological capabilities. (AI) technology can change the nature of work and employment is rapidly changing to make the best use of both humans and (AI) talent in healthcare industry in the future. For example, (AI) technology offers a way to fill in gaps and the rising labor shortage in healthcare. According to Accenture analysis, the physicians shortage is increasing. However, (AI) technology will manufacture healthcare machine men to replace physicians in future one day(2017). Hence, (AI) technology will be invented to raise health care service staffs work efficiency and performance in any hospitals or clinics in the future.

In conclusion, (AI) technology will raise efficiency for any service or manufacturing industries in the future, although, it is possible that it will also rise low skillful workers unemployment numbers. But, the most important influence to human technological innovation will be risen and it will influence human life will be changed to be better, e.g. self drive cars, health care physician machine men, machine man cleaners etc. intelligent machine men will be manufactured to serve for our daily life. Furthermore, (AI) technological products will influence countries trading, some low technological development countries manufacturing businessmen can choose to buy any (AI) products to raise whose productivity and efficiency and reducing cost to achieve economic cost saving result. Also, GDP of trading growth income will increase to the (AI) products sale countries. Hence, it will be beneficial to economic development to both developed and developing countries both in the future as well as (AI) scientists time and money spending will be valued to continue to invest (AI) technology development for human life and economy benefits for long term.

In conclusion (AI) technology will raise macro economy growth and it can create many (AI) jobs , but it also raise the low level technological worker unemployment change. In the future, (AI) technology can be applied to digital technology to attempt to invent any new undiscovered (AI) and digital technology. So, it needs any scientists to continue to research how digital and (AI) technology can be mixed to satisfy human's future undiscovered needs.

# FOUR

# How Robots Change Job Nature

Although, (AI) technology will be popular to applied to different jobs, but it still needs social acceptance to replace some human jobs. Today, it is increasingly common for people to use robots in various situations at home and in retail stores, hotels and hospitals. Robots are classified into several types based on their functionality ( service and utility robots or those designed to communicate with humans ) and appearance ( humanoid robots or mechanical robots). The types of robot to which every country attaches particular important in the advance of robotics, reflects the sense of values and preferences of its population . Thus, (AI) will be applied to replace human to do these above different kinds of job nature. For example, U.S. has the highest level of robot utilization at home and an retail stores with its people being the most enthusiastic about the future use of robots. Otherwise, Germany shows a strong tendency to consider robots for industrial purposes, and its people feel strong to the presence of robots in their households. Japanese accepts to apply" human aid robot" that can communicate with humans and they have a high level of familiarity with robots.

Hence, it implied those three countries have accept (AI) to replace human to do any these kinds of job duty and it will influence these three countries' workers lose their old occupations and who will unemployed absolutely, due to many (AI) robots replace them to do their job duties in the future. Also, US will have many retail service workers or retail warehouse workers are unemployed. Germany will have many manufacturing industry's workers are unemployed. Japanese will have many communication industry workers are unemployed, such as telephone service, shopping center services etc. different kind of service industry's service staffs . It will cause these kind of workers' competitive abilities are lost in themselves countries' jobs that require such skills include software developers, court judges, nurses, high school teachers, dentists and university lecturers, these occupations are still difficult to be replaced by (AI) robots.

Are robots taking our jobs or making them? In fact, our societies will have unemployment challenges, even (AI) technology has not created before. However, after (AI) robots invention, some of human jobs will be replaced and it can raise many low skillful and low knowledge level worker unemployment number. However, I think that high productivity driven by increasingly powerful IT -enabled machines is the causes of global labor market problems and accelerating technological change will only make those problems worse.

IT technology brings this question: Are robots killing human's jobs or benefiting human's jobs? I suppose that there is a limited amount of labor to be done. The implication is that technology can create unemployment by displacing workers, such as (AI) invention, because the more efficiently worker work ( using machines or (AI) robots), the loss work there is for workers to do. Even, any new jobs will be better done by machines or (AI) robots, and unemployment will still skyrocket. How do we know that humans will always be better at some work, or more importantly, enough work, than machines or (AI) robots, e.g. human drivers drive more safe or careful to compare (AI) robot drivers. But, the challenge is that it is not ensure that (AI) robots drivers must not drive careless to cause the chance of accident occurrences more than human drivers. However, technological change can be beneficial to innovation, automation and increasing productivity for businesses.

Consequently , it may seem machines can hurt wages and job for low skillful, less educated workers. Also, high educated workers are likely as less educated workers to find themselves displaced and devalued, and more education may create as many problems as it solves. Thus, in negative influence, automation effects on particular jobs shift workers to other jobs that are equally or more desirable. Workers may be highly compensated for possessing human capital that is specialized to a labor market. If technological advance is very rapid, such as (AI) invention, causing a large and very rapid drop in demand in a large labor market, the economy may not be able to absorb the sudden surplus of labor in a short period of timer when (AI) robots are popular to replace some workers to do some occupations in global societies.

For example, self-driving vehicles threaten to send truck drivers to the unemployment office. Computer programs can now write journalistic accounts of sporting events and stock price movement. There are even computers that can grade essay revolutionize some part of teaching jobs. Hence, (AI) robots will have possible to replace human brain to do any judgement, argument, and mind job duties. It implies some occupations which need human' mind will be threaten by (AI) robots, e.g. author, accountant, nurse, engineer. Thus, (AI) robots will have possible to replace some professional and high educated workers' jobs in the future.

But, technology can create new nature of jobs in possible. For example, a 60 minutes program indicated technology is putting new categories of jobs in the sites ( sic) of automation, the 60% of the workforce that makes its living gathering and analyzing information. Also, recession: technology kills middle -class jobs that overall technology is eliminating for more jobs than it is creating by (AI) technology. Hence, human's brain work may be assisted by 60% of (AI) gathering and analyzing information for some occupation , e.g. space scientists, ocean scientists, earth scientists etc.

However, I believe the (AI) invention and human job competition may influence global productivity change. Productivity is economic output per unit of input, the unit of on input can be labor hours( labor productivity), but if (AI) robots replace human job, then the unit of input may be (AI) machine hours ( AI) robot productivity or all production factors including labors, machines and energy ( total factor of productivity). Producing more output with less input can take several forms.

The traditional notion of productivity is a form reorganizing production and/or using better or more technology to produce more output per worker hour. But when (AI) robots invention, the form can be reorganizing production and/or using better or more (AI) robots to produce more output per (AI) robot hour. Hence, if the firm apply (AI) robots to produce its products. Then , productivity improvements in the firm may result in less workers employment, due to (AI) robots replace more worker number to achieve more productivity improvement, it has economic benefits ( less factor of production) , but more production in long term.

Thus, (AI) robots can help any firm to achieve productivity improvement in long term, for example, if unproductve farmers move to the city and start working for high-tech. manufacturers. The shift effect can be more dynamic and disruptive as low-productivity industries lose out in the marketplace to high -productivity industries and the compositional mix of the economy changes. Thus, in the long term (AI) robots can also be beneficial to high productivity industries to bring the mix of economy positive changes.

Moreover, automation will also produce some new jobs in firms that sell the new robot or other labor-saving technology. This means that, in general, there will be shift in the economy in the direction of higher-skill and higher wage jobs. Even if the (AI) robot invention country, US becomes a leader in (AI) robots producing productivity-enhancing technology, it will experience a growth in jobs serving foreign (AI) robots product buyers. Hence, (AI) robots can also create (AI) salespeople, (AI) manufacturing workers , (AI) inventors, scientists, (AI) software designer etc. occupations, when if all society does is move workers from insurance firms, restaurants and car factories to robot factories, productivity will have remained the same to create job needs for insurance, restaurant and car manufacturing worker service occupations for (AI) software designer, (AI) service robots manufacturer, (AI) service robot seller etc. related (AI) service robot product occupation created in (AI) robot technology job market. Hence, (AI) invention also create new (AI) technology job chance. (AI) impacts management job market.

In future, organization management will be changed from (AI) introduction. Division of labor will change and collaboration among humans and machines will increase. Companies will have to adapt their training, performance and talent acquisition strategies to account for a new found emphasis on work that hinges on human
judgement and skills, including experimentation and colloboration.

How (AI) impacts any organizational administrative management work? (AI) 's greatest impact will be on administrative coordination and control tasks, such as scheduling, resource allocation and reporting, (AI)-driven will place a higher premium on what we call " judgement work", the application of human experience and expertise to critical business decisions and practices when information available is insufficient to suggest a successful course of action. This kind of work will require new skills and mindsets; replacing people with machines is not goal in itself. When, artificial intelligence enables cost-cutting automation of routine work, it also empowers value -adding augmentation of human capabilities.

Thus, administrative and routine tasks, such as scheduling, allocation of resources, and reporting, will within intelligent machines, responsibilities that have long been reserved for humans. For instance, a typical store manager or a lead nurse at a nursing home must constantly juggle shift schedules, accounting for staff members' absense owing to illness, vaction, time or sudden departures. Many of these tasks will be automated by (AI). Imagine (AI) writing management monthly reports, it is not a distant dream. Leading news providers and Wall street banks are now using (AI) report generators to write news and analytical reports by drawing on quantitative data. The associated press, for example, expanded its quarterly earnings reporting from approximately 300 companies to nearly 3,000 with the help of (AI) powered software robots, freeing up journalists to conduct more investigative and interpretive reporting. For another example, Jobalime, a job-placement site, uses intelligent voile analysis algorithms to evaluate job applicants. The algorithm assesses paralinguistic elements of speech, such as tone and inflection, products which emotions a specific voice will elicit, and identifies the type of work at which an applicant will likely excel. In the future , (AI) machines can be applied to assist some kind of office administrative jobs duties. It's attractive to office managers to achieve more accurate judgment to do any administrative matters when who can be assisted from (AI) machines. Thus, managers need to spend time to learn how to apply (AI) machine to assist them to do more accurate judgement, and better informed choices. (AI) robots can be applied to improve the speed quality and cost of available products and services, instead of applying on productivity improvement and administrative improvement aspects. Thus, they may also displace large numbers of workers. This, possibility challenges the traditional benefits model of trying health care and retirement savings to jobs.

In an economy that employs dramatically fewer workers to deliver benefits to displaced workers. For example, the worldwide number of industrial robots has increased rapidly over the past few years. The fall prices of robots, which can operate all day without interruption, make them cost- competitive with human workers. In special consideration, in the service sector, computer algorithums can execute stock trades in a fraction of a second, much faster than any human. As those technologies become cheaper, more capable, and more widespread, they will find even more applicants in an economy.

Consequently, (AI) technology brings unemployed number increasing many businesses continued automating their operations rather than hiring additional workers. A trend among technology companies that receive massive valuations with relatively few workers. For example, in 2014 year Google was valued at $370 billion with only 55,000 employees, a tenth the size of AT & T's workforce in the 1960 year. Hence, if automation technologies like robots and artificial intelligence make jobs less secure in the future, there needs to be a way to deliver benefits outside of employment " flexi security" or flexible security is one idea for providing healthcare, education and housing assistance whether or not someone is formally employed.

In conclusion, (AI) and robots technology will raise unemployment to some occupations when (AI) replaces same industries' workers job duties in our societies in the future, but it also create new jobs to raise employment in any related (AI) robots and automated machine products in (AI) manufacturing. (AI) design, (AI) sale self-related industry, when (AI) replaces same industries' workers' job duties.

(AI) journalism, media publishing, digital
communication technology trend

How to apply (AI) technology in digital communication journalism media, publishing industry? Some scientists indicate future (AI) and digital technology may consist such as: voice driven assistants, emerge. For example, Amazon e book publish applying digital technology and (AI) auto printing technology to sell e books to let readers to listen any e book content by (AI) voice driven speaker when they turn on computer to read e book contents; capable phones start to unlock the possibilities of 3D image of mobile story telling. New smart wearables include ear buds that handle instant translation and glasses that talk and hear. China and India will become a key focus for digital growth with innovations around payment online identity, and artificial intelligence. Thus, future (AI) technology can be applied to 3D image mobile story telling, online payment method to dealt online transaction publishing industry.

Thus, future (AI) technology can be applied to online e book publishing industry to make sound books to let readers feel more attractive . Such as Amazon publish has published sound e books to attract readers to choose to read any its books from online. Also, (AI) technology can also be applied to communication industry. For example, some online pure-play news, opinion and entertainment websites. It is a digital communication media, e.g. online journalism blog (AI) technology can be applied to visual storytellers to let online book readers to enjoy to listen to watch and send any online electronic book contents more attractive. Thus, future (AI) technology will be popular to assist any electronic book publishers to publish visual and sound talking storybook to let readers who can watch motive image and listen and read words from e books more attractive.

Thus, (AI) technology can be applied to internet ecommerce publishing or media industry to help any electronic book publishers to publish sound, image motion electronic book to attract global readers to read, even (AI) technology can be applied to digital entertainment industry, e.g. electronic 3D image virtual video games, computer games. It can be also applied to education industry, e.g. the first true digital native generation and are the native speakers of the digital language of computers to let student to learn different languages or translate words to compare to classroom learning more easily. It can be also applied to communication industry, e.g. (AI) mobile phone. Hence, it seems (AI) technology can be applied to publishing, communication, education , entertainment etc. different industries in the future. (AI) technology will be one kind of tool to satisfy human's daily life needs in the future and these industries has one characteristics is that they need to apply internet to operate to operate to do online business.

Thus, it has three trends of (AI) technology and internet technology need to be linked to achieve one kind of attractive technological business to satisfy client's needs. These three trends as below: All consumer trends involve the internet. It will be many consumer's online habits, shopping, working, socializing, watching TV, studying, travelling, listening. Thus, (AI) music, eating and exercising are just a few examples. This is happening because human usually use mobile broadband or Wi-Fi, rather than cables. Thus, (AI) technology will be applied to mobile to satisfy client's need absolutely.

The mobile phone can be more popular to be used more than computer or laptop tools. The reasons are because women dive the smartphone market by defining mass-market use. But as the speed of technology adoption increases mass market use becomes much quicker then before. Successful new technological products and services , such a (AI) mobile phone products now reach the mass market in popular use. It means that the time period when early adopters influence others is shorter than before. Also, since new products and services increasingly use the internet mass markets are not only faster , but are also more important than ever to consumer themselves. Most internet services become more valuable to individuals when many use them. Thus, it causes why (AI) mobile phone will be popular to be used.

Since, new products and services increasingly use the internet, in the future several trends focus on (AI) smart phone users. Consumers' familiarity with using smartphone apps. Essentially, the technologies will bring other related (AI) and internet service needs, e.g. sound and image emotion e book needs, (AI) mobile communication needs, e-virtual games or e-3D image virtual games etc. entertainment activities needs with such a large part of the world's population now online, it is clear that there is strength in numbers.

Thus, (AI) imagines , if future any (AI) and internet related services or products new technology is easy to use and inexpensive, when the latest products reach the mass market almost as quickly as they reach the early adopters and industry experts. I believe that any (AI) and internet related products or services must be popular to accept to consume for entertainment or useful aim. For example, with major players including Apply, Facebook and Google had invested (AI) technology to develop their businesses. (AI) technology has the potential to disrupt everything in the coming years, from the lives of connected consumers to every industry (AI) will be an alternative route for brands to reach consumers with convincing and relevant messages. Digital technology will assist of the future, then it can improve technology to bring this effect, such as sophisticated software machine learning and speech recognition effective. Hence, Google, Facebook , Yahoo web site service companies can apply (AI) technology to help other companies to advertise their businesses, such as travel, retail, and education etc. industries more attractive. (AI) technology can be applied to internet company to be aware and familiar enough to drive among mainstream consumers, it can create online experience to travel, retail , education and other entertainment needs to online consumers to seek their entertainment needs more easily. Hence, in the future (AI) technology and internet related entertainment service needs will be raised in this (AI) and online consumption market.

● (AI) healthcare service industry development

In the future, (AI) medical internet technology tool can be applied to assist individual's health at the center of their focus, e.g. smartwatch compatible mobile app. patients can let personalized reminders for taking their

medication snap pictures of their prescriptions to expedite refills, and scan their insurance card. So that, store clerks are prepared with up-to-date patients' information . (AI) owned health operated technological clinics can help patients to receive treatment for minor illnesses, flu shots, cholesterol screenings and more than a dozen other medical services, all of which can be patients who can't make it to a physical location. (AI) healthcare services organizations can provide various telemedicine services. So, patients can receive care via phone or video chat.
For example, one London-based intelligent Brewing company has developed an (AI) system to continuously collect and incorporate customer feedback, which the system itself uses to brew ne various of the company's beers. Thus, the beer clients can give feedback to talk to the algorithm (AI) machine, whenever or anywhere who're drinking the beer. It is such any healthcare services organizations can apply (AI) machine to collect patient's feedback to talk to the algorithum (AI) machine whenever or anywhere who're eating any medicines. So doctors can know every patient's health conditions any time. If the patients feel uncomfortable, the doctor can know from (AI) machine notification to decide whether the patient needs to eat another new medicine or keep to eat same medicine is better. Hence, (AI) medial internet technological body check report machine will be proper to be needed to serve any hospitals' patients in the future.

However , it brings this question. How can (AI) medical internet technological body check report machine apply to hospital more efficient? The essential new medicine co-workers for the health service digital age health service leaders need apply (AI) medical report machines and artificial intelligence to the newest recruits to the workforce bringing new skills to help health service staffs do new jobs and reinventing what's possible, building the health service workforce for today's digital health service demands for patients. Thus, technology-driven health service model innovation from the health service organization outside in and providing digital health service ecosystems for patients to use the (AI) health service equipment will be popular to be accepted to be used.

- I Robot and internet things future machine
men invention

Nowadays, there are some company, which apply internet and (AI) I Robot technology to do any similar human job nature. For fishing industry example, one company, known for creating the Roomba, I Robot is now working with marine conservationists to launch an ocean-patrolling intelligent robot to hunt and manage invasive species, protecting native populations. And evolved industries like precision agriculture are ramping of our increasing population. Area of practice that once seemed impossible to digitize are fundamentally changing because of the impacts of (AI), internet of things capabilities and big data analytics, which have many potentially positive impactions for society.

For textile industry example, automation is nothing new, it has shaped the workplace to replace human jobs to boost productivity in the textile industry. Textile machines have had a generally positive impact over gears, creating value and allowing textile workers to take up more rewarding age will likely continue to create opportunities and lead to new textile industries, companies and textile occupations. It may also compensate for a demographically driven slowdown in the growth of the textile workforce. The future impact of textile (AI) and automatic and internet link is somewhat uncertain. It seems textile industry will be trend to accept (AI) textile workers and internet of thing to replace traditional manual textile workers to produce any shirts, cloths etc. wearing products in factories popularly, during the (AI) textile machine and internet thing technology can be invented to reach the mature stage in the future.

For factory worker transportation job example, they have also expanded their influence, migrating from the factory floor to the service sector and taking the place of humans in a range of activities from financial transactions to transport route optimization. Further (AI) machines and robots are increasingly programmed to learn, meaning they improve with time and undertake cognitive activities. Hence, (AI) machines and internet technology enable automation of work activities to raise factory workers' efficient and performances, also factories can reduce manual worker numbers, due to (AI) machine workers' assistance.

Future, (AI) robotics technologies and internet technique have these different kinds of characteristics: For soft robotics example, it is non-rigid robots construct with soft and deformable materials that can manipulate items of varying size, shape and weight with a single device. For swarm robotics, it coordinated multi-robot systems often involving large numbers of mostly physical robots. For touch/factile robotic example, it robotic body pails (often biologically inspired hands) with capability to sense, touch , dexterity robots example, serpentine robots with many internal degrees of freedom to threat through tightly packed spaces for humanoid robots example, robots physical is similar to human being often bi-pedal that investigate variety capable of performing human tasks , including movement across terrains, object recognition, speech sensing etc. For autonomous cars and trucks example, it is capable of operating with a human pilot, e.g. the unarmed general atomics Predator XPUAV with roughly half the wingspan of a Boeing 737 can fly autonomously for up to 35 hours from take-off to landing, for unmanned aerial vehicles example, flying vehicles capable of operating without a human pilot, the unarmed general atomics predator -XPUAV , with roughly half the wingspan of a Boing 737, and fly autonomously for up to 35 hours from take off to landing, for (AI) chat bots example, (AI) systems designed to simulate conversation with human users, particularly those integrated into massaging apps.

In Dec. 2015. the general service administration of the US Govt. described how it used a chat bot named Mrs. Landingham (a character from the television show the west wing) to help onboard new employees. Finally, for robotic process automation example, class of software robots that replicates the actions of a human being interacting with the user interfaces of both software systems. Enables the automation of many back-office work flows without requiring expensive IT integration . Hence, future (AI) robot machine men will have different functions to be applied to different industries to use in possible.

Statistics Denmark shows that (AI) automation potential robots will influence few jobs are completely automatable , but close to half consists of 40% automatable tasks: It showed example occupations include share of automated, such as brewing machine operators are more than 80%, logging equipment operators are more than 50%, roofers , stock tasks clerks, travel agent are more than 50%, farmers , nursing assistants are more than 30%, physicians, teachers , managers are more than 10%.

For example, humans perform a wide variety of tasks from planting corn to examine spreadsheets, meeting clients and lifting crates in a store. Each of these actions requires a combination of innate or acquired capabilities, internet technique assistance, ranging from social perceptiveness to fine motor skills and natural language understanding. To understand and map automation feasibility by existing technology. Mc Kinsey has developed a framework of 18 technical capabilities that can substitute tasks performed by humans. The capabilities are grouped in five categories: sensory, cognitive, language, social and emotional and physical. So, it seems (AI) robot machine men and internet technique will have possible combination to invent to own human's emotion , language, learning, task skill abilities.

Mckinsey global institute analysis also showed current technologies have achieved different levels of human performance across 18 capabilities include: sensory perception, autonomously infer and integrate complex input using sensors, cognitive capabilities reorganizing known patterns/ categories supervised learnings, generating novel, logical reasoning/ problem solving, optimization and planning, creative, information retrieval, coordination with multiple agents, output articulation/presentation, national language processing, social and emotional capabilities-natural language understanding, social and emotion sense, reasoning output, physical capabilities-fine motor skills, navigation mobility. Hence, it seems (AI) robots and internet technological will combine to invent to own human' some skills to replace human to do some kind of tasks in possible.

In conclusion, future (AI) robot and internet will be needed to link to cooperate together to raise human's work efficiency in popular.

How artificial intelligence replaces human job possibility

What is the risk of automation for jobs to replace human job? In recent years, there has been a revival of concerns that automation and digitalization night after all result in jobless future. As I argue, this might lead to an overestimation of job (AI) automate , as occupations labelled as high-risk occupations often still contain a substantial share of tasks that are hard to automate.

For example, when the share of (AI) automatable jobs is 6% in Korea, the corresponding share is 12% in Australia. Differences between countries may reflect general differences in workplace organization, differences in previous investments into (AI) automation technologies as well as differences in the education of workers across countries. I also discover that (AI) automation and digitalization are unlikely to destroy large numbers of jobs. But, however, low qualified labors are likely to raise costs as the (AI) automate of their jobs is higher compared to highly qualified workers.

In fact, (AI) technology will influence some new technology to replace some human's job, such as driverless car, the largely autonomous smart factory , service robots or 3D printing. These technologies are driven by advances in computing power, robotics and artificial intelligence and ultimately redefine what type of human capabilities machines are able to do.

Hence, question brings whether (AI) invention will influence general human jobs to be replaced by (AI) autonomous jobs? Whether will the potential foe automation with actual employment loss? In particular, the technical possibility to use (AI) machines rather tasks need not mean that the substitution of humans by machines actually takes place.

Whether (AI) technology replaces human's some job, it is beneficial to our society or not. Instead, machines are increasingly capable of performing non-routine cognitive tasks, such as driving or legal writing . In particular, advances in the field of machine learning (ML), e.g. computational statistics and visions, data mining, artificial intelligences allow for automating cognitive task, when the use of (ML) in mobile robotics (MR) also allows for automating certain manual tasks. So, it seems, (AI) technology can replace some labor job, e.g. warehouse transportation, even mind's job, e.g. legal writing, driving in possible.

For example, if (AI) automatic non-manual driving can reduce hurt or death risk, it is beneficial to our society, or (AI) automatic robots can more any heavy things ( products) in warehouse safely. Then, it can reduce the warehouse labor's bodies hour risk, it is beneficial to the workers. Even, if (AI) robot can write any legal documents, no any word errors in short time. It is beneficial to the law companies , but it also bring unemployment chance, due to these jobs can be replaced by (AI) robots to do in the future. Hence, it will cause some occupation to be disappeared, due to (AI) robots can do our these kinds of jobs in the future.

Frey & Osborne (2013) reported these kinds of occupations will be replaced by (AI) robots in possible. They include computer, engineering, financial, management, legal , art and medium, community service, education, healthcare practitioners and technical service, sales and related, office and administrative support, farming, fishing and forestry, construction and extraction, installation, maintenance, and repair , production, transportation and material moving. It seems our future some professional occupations will have possible to the replaced by (AI) robots to replace, instead of labor jobs. Hence, (AI) robots technology will have much trend to replace high knowledge or low knowledge skillful labors in the future.

In conclusion, it implies that only using information on task-usage at the individual level leads to significantly lower estimates of jobs " at risk", some workers in occupations with according to high automate nevertheless often perform tasks with are hard to automate. Why can (AI) replace human to do some kinds of jobs? (AI) artificial intelligence refers to the ability of a computer or a computer enable robotic system to process information and produce outcomes in a manner similar to the thought process of humans in learning, decision making and solving problem. By extension, the goal of (AI) systems is to develop systems to capable of tasking complex problems in ways similar to human's logic and reasons who feel in our future. Hence, it means future (AI) robots has effort to replace human to do any jobs in possible.

(AI) directions for future non-manual
control road vehicles market

Future road vehicle products and technologies must meet social, economic and environmental protection and driving safety goals , and satisfying market requirements for mobility, accident reducing, performance, cost desirability. Thus, (AI) auto-non manual control vehicles need to be followed this direction to invent. To satisfy future driver's safety of needs, enhanced vehicle speed desired functional performance of road transportation system, required and desired technological response, including research needs. It is long term up to 20 years vision, for (AI) auto non-manual research. Thus, (AI) auto non manual vehicle manufacturers need often to revise their (AI) vehicles functions to raise to improve their system performance and driving industry driver's needs, e.g. private drivers need or public transportation driver's need or business client's need. Hence, future (AI) transportation will need have individual driving consumer and business driving consumer both targets.

Thus, future (AI) automation manual control vehicles need to deliver high impact technology solutions to meet social , economic and environmental and safe goals. Engine needs to be improved efficiency , performance, drivability, reliability, durability and speed-to-market together with reduced emissions and cost; hybrid, electric and alternatively fuel (AI) non manual control vehicle technology development, leading to new fuel and power systems, such as hydrogen, fuel cells and batteries, which satisfy future social, economic and environmental and safe goals. Software, sensors, electronics and telematics technology development are needed to be lead to improve vehicle performance, control and adaptability, intelligent , mobility and security, structure and materials technology development, leading to improved safety, performance and leading to flexibility with reduced cost and environmental pollution to achieve the (AI) non manual drivers to feel (AI) vehicle performance, auto control and adaptability is better to compare traditional manual driving vehicles.

In fact, in traditional manual driving market, Japan and USA had had over 80% of world car production by six major global groups. In the future, it is possible only USA can dominate (AI) non manual auto driving vehicle manufacturing market if Japan had no effort to manufacture any (AI) auto non manual control vehicles. So, it means that it is only Japan is USA potential (AI) auto non manual control vehicle manufacturing competitors. Also, it means that it is only USA has effort to export (AI) auto non manual control vehicles to global (AI) auto non manual control vehicle market.

Thus, in long term, (AI) non manual control vehicle product market development, USA (AI) vehicle manufacturers will have these requirement to win new technological competition to traditional manual control vehicle. The requirements include: low cost fuel, low carbon, fuel cell and telematics technologies, the technological roadmap function, such as detailed consideration clear provision of other important areas to the drivers. When the (AI) non manual auto drivers are sitting in the non manual control auto vehicle. Although, who does not need to drive, but who need to know how to go to anywhere by electric road map show clearly. So, the driver won't lose direction and he/she can know the (AI) non manual control vehicles is driving to anywhere in any time, even when who is sleeping.

In the future, the (AI) non manual control vehicles need to be invented to satisfy any business , transportation clients' needs, instead of individual clients needs, e.g. cans, trucks, buses, emergency and utility vehicles, trains, trams etc. Hence, technological road mapping is one important tool to help any business, transportation ( AI) non manual control vehicle clients. Technological roadmap is a technique that is used in industry to support strategic planning for (AI) non manual driving vehicles in the future. Electronic road maps generally take the form of multi-layered time based charts, linking technology developments to future (AI) non manual control vehicle market requirements.

Technology road mapping is a flexible technique and the roadmap architecture and process for developing the roadmap most generally be customized to meeting the particular aims. Why technology roadmap will be popular to (AI_ non manual control vehicles. It's advantages include: It is a technology solutions and options that can enable the performance targets to be achieved engine hybrid, electric and alternatively fueled vehicles, software, sensors electronics and telematics, structures and materials design and manufacturing process. It is road transport system performance measures and targets tool, in response to the trends and get (AI) non manual control vehicle drivers to get society, economy, environment protection, low cost driving benefits, also it can help any transportation clients to know how to go to anywhere clearly. Hence, technological road map will be one good tool to assist (AI) non manual control auto vehicle to develop future road driving market.

Reference(source)

Frey & Osborne (2013), The future of employment: How susceptible are jobs to computerization? University of Oxford.

Mckinsey Global Institute Analysis

Statistics Denmark, Global automation impact model, Makinsey analysis

(AI) development second stage

(AI) -driven automation industry development

(AI) -driven automation industry will create wealth and expand economy growth to any countries, but it will be accompanied by changed in the skills that workers need to learn. One of main ways that technology increases productivity is by decreasing the number of labor hours needed to create a unit of output. It implies (AI) technology will influence low educated and low skillful labor number to be decreased ( reduction employment number). In contrast, technological change tended to work in a different direction throughout the nowadays. The advance of computer and the internet raised the relative productivity of higher skilled workers. So, routine-intensive occupations that focused on predictable tasks disappearance, such as switch board, operators, filming checkers, travel agents and assembling line workers etc. were particularly replaced by new technologies.

However, today, it may be challenging to predict exactly which jobs will be most immediately affected by (AI) driven-automation. The reason is because (AI) is not a single technology, but rather a collection of technologies that are felt unevenly through the economy to influence job changing both negatively and positively. In positively view point, (AI) driven-automation will make many workers more productive and increase demand for certain skills. Consequently, new jobs are likely to be directly create in areas , such as the development and supervision of (AI) as well as indirectly created in a range of areas throughout the economy as higher incomes lead to expanded demand. Otherwise, in negatively view point, many traditional human needed ( demand) skillful jobs will be threatened by automation are highly concentrated among lower-paid, lower-skilled and less -educated workers. It means automation will cause pressure on demand for this group, pressure and employment, if (AI) can replace the low skilled and less educated workers' jobs. Thus, (AI) will have negative influence to impact on the labor market.

(AI) capabilities will enable automation of some tasks that have long required human labor. Why can (AI) replace some simple human jobs? For example, advances in robotics are expanding machines' abilities to interact with and sharp the physical world. Combined , (AI) and robotics will give rise to smarter machines that can perform more sophisticated functions than ever before and brings more advantages that humans have exercised. This will permit automation of many tasks now performed by human workers and could change the shape of the labor market and human activity.

How (AI) influences labor market

Today, it may be challenging to predict exactly which jobs will be most immediately affected by (AI)-driven automation. Because (AI) is not a single technology, but rather a collection of technologies that are applied to specific tasks.

Some specific predictions are possible based on the current (AI) technology. For example, driving jobs and house cleaning jobs, bank counter service jobs, telephone enquiry service operators. Restaurant cooking jobs, simple accounting record service jobs etc. that require relatively less education to perform. Advancements in computer vision and related technologies have made the feasibility of fully appear more likely, potentially displacing some workers in driving-dominant professions. Seemingly similar robot, for which the operational tasks is less specific of navigating to a specific destination when following a set of given rules and preserving safety.

In the future, the effects of (AI) on the labor market in the decade ahead will continue the trend toward skill-biased change that computerization and communication innovations have driven in recent decades. Thus, some human driving occupation will be disappeared or replaced by (AI) automation driven. For example, bus drivers, light truck or delivery services drivers, heavy and tractor-trailer truck drivers, school drivers, tax drivers, travel bus drivers.

However, (AI) technology could enable some workers to focus time on other job responsibilities, boosting their productivity, and actually raised wage

growth among those still holding the reshaped jobs. For example, salespeople, who currently spend a considerable amount of time driving could find themselves able to do other work when a car drives them from place to place, or inspectors and appraisers could fill out paperwork, when their car drives itself. This (AI) -driven technology should make these workers more productive, with (AI) -driven technology serving as a complement, not a substitute. New jobs will also likely be created, both in existing occupations cheaper transportation costs with lower prices and increase demand for products and all the related occupations, such as service and fulfillment, and in new occupations not currently foreseeable.
What kind of jobs will be created by (AI) technology? Predicting future job growth is extremely difficult, due to it depends on technologies or substitute for existing today as well as they may complement or substitute for existing human skills and jobs. However, (AI) will also lead to substantial indirect job creation to the degree it raises productivity and wages, it may also lead to higher consumption that would support additional jobs from high-end draft production to restaurant and retail. The future(AI) " augmented intelligence", the technology's role is as assisting and expanding the productivity of individuals rather than replacing human work. Thus, based on the biased-technical change framework, demand for labor will likely increase the most in the areas where humans complement (AI) automation technologies. For example, (AI) technology , such as IBM's Watson may improve early detection of some cancers or other illnesses, but a human healthcare professional is needed to work with patients to understand and translate patients' symptoms, inform patients of treatment options, and guide patients through treatment plans. Shipping companies may also partner workers who pick up and deliver products over the last feet with (AI) enabled autonomous vehicles that move workers efficiently from site to site. In such cases, (AI) augments what a human is able to do and allows individuals to either be move effective in their specially task or to operate on a larger scale. Thus, it seems (AI) technology will also create new jobs, raise productivities and workers' efficiencies.

Redefining management in the workforce of artificial intelligence
In the future, due to artificial intelligence influences to some kind of human jobs nature. So, the kind of human jobs of management methods will also need to change to adapt the artificial intelligence technology input to their organizations. It will cause challenges for every executive and manager

if who won't have effort to manage their teams how to apply artificial intelligence technology to work efficiently and easily. For example, division of labor will change among humans and machines will increase. Thus, companies will have to adapt their training performance and talent strategies how to emphasize on work that how to make human judgment and skills and experimentation. Thus, (IA)'s greatest impact will be on administrative coordination and control tasks, such as scheduling , resource allocation.

In fact, mangers will encounter this challenges: How to apply human experience and expertise to judge critical business decisions and practices when the information available is insufficient to suggest a successful course of action? Due to this kind of work will require new skills and mindsets. I shall indicate these change management methods to adapt (AI) technology. Such as: administration and routine tasks, scheduling , allocation of resources and reporting will fall within the intelligence machines, responsibilities that have long been reserved for humans. For example, a typical store manager or a lead nurse at a nursing home most constantly arrange shift schedules, accounting for staff members' absences owing to illness, vacation time or sudden departures.

Thus, the managers need to learn how to arrange new division of labor within the organizations after (AI) technology had been implemented to the organization. Artificial intelligence is currently influencing into once considered exclusive to humans: assessing and acting on human emotions and personality traits. The influences to managers need to change their strategies to adapt (AI) technology implements include such as below:

Firstly, managers need to spend the bulk of their time on coordination and control tasks from intelligent system implements. Their time spending on these major three aspects from impact of intelligent system: coordinate and control, solve problems and collaborate and people and community , strategy and innovation three aspects. Thus (AI) will influence managers need to change their judgment method to teach whose teams how to adapt the (AI) system operations in any organizations.

Secondly, (AI) will influence top, middle and low level management needs to change to adapt the (AI) technology operations to any owned (AI) technology organizations in the future. Intelligent machines must be trained in context. Just like humans , on-the-job training is a requirement for such machines because they typically arrive with only very general capabilities. To get the most from (AI), managers at all levels must participate in the instructional experience and in the learning process and provides managers' familiarity with such systems on these aspects, e.g. How the system works and generate advice, how the system has a proven track record , how the system provides convincing explanations , how the system can make simple rule- based decisions.

Thirdly, managers need to learn how to make judgment more accurate (AI) systems assistance. Although (AI) will invariably take on more routine work and even augment human decision-making, it won't judgment work, the application of human experience and expertise to critical business decisions when the information available is insufficient to suggest a successful course of action or reliable enough to suggest an obvious course of action. For a sense of the nature of judgment work, consider big data marketing and sales analytics. Such analytics often provide insights that can inform promotional campaigns, including predicting which promotions will generate desired sales brand further into the future, marketing executives need use judgment, combining analytics with their own and others' insight and experience.

The application of experience and expertise to critical business decisions and practice represents the real value of human judgment. But, when artificial intelligent machines are implemented to any organizations to assist the low, middle and top level management to make any business judgment. These forms of judgment work that managers can gather data interpretation, idea development more absolute from (AI) machine assistance. Thus, why these level management executives need to learn how to apply (AI) machines to help them to make any business judgment more accurate.

How (AI) influences organizational change

Consequently creative and social intelligence will be in even greater demand as (AI) makes in management and the workforce. This development will represent a long term trend in labor markets , one characterized by intensifying demand and reward for social skills with a growing desire for creative capabilities, managers will seek to fashion of ideas and hypotheses from inside and outside of the enterprise to shape solutions to their most pressing business problems. Thus, (AI) will influence overall organizational team members who have chance to participate any decision to make more accurate business judgment.

Many managers mistakenly view judgment work as only an individual discipline, failing to appreciate that it can also involve decide interpersonal and organizational practices. In more complex settings, judgment is typically a collective outcome of individuals‘ and teams’ diverse perspectives, insights and experiences. And often , the resulting choices are better informed than decisions that an individual would have arrived at on his or her own.

Thus, when any organizations apply (AI) technology to assist managers to gather data and ideas to make any judgment. In these cases, organizations can create the conditions for effective collective judgment by establishing structures , such as " shadow advisory boards" that prompt managers and employees to source and synthesize multiple perspectives. Thus, a traditional organization (firm) might freshen its thinking is t put together a shadow advisory board, comprised of young, digital people who can apply (AI) machine assistance to make judgment work more accurate whether related to people development, problem-solving or strategizing and innovating for considerable degrees of creative and social intelligence.

Thus, on the one hand, (AI) technology machine augmentation and automation can give these advantages to human (organization managers) , e.g. developing people and community, solving problems and collaborating, coordinating and controlling work, shaping strategy and leading innovation. Besides, on the other hand, the next generation managers need have these individual attitude to treat intelligent machines to be as colleagues.

When, judgment is a human skill, intelligent machines can accelerate human learning that supports it, assisting in data -driven simulations, scenarios and search and discovery activities. Focuses on judgment work, some decisions require insight beyond what data can tell them. This is the sweet sport for human judgment, the application of experience and expertise to critical business decisions and practices. Thus, managers will also need to find ways to learn how to use digital (AI) technologies to tap into the knowledge and judgment of partners, customer external stakeholders and role models in other industries after the (AI) machine had been implemented to the organization.

Future works change:
Automation, employment
and productivity

Human future " micro to macro" industry trends will be affected business strategy and public policy by (AI) technology. In the future (AI) technology will influence those six themes: productivity and growth, natural resources, labor markets, the evolution of global financial markets, the economic impact of technology and innovation and urbanization. However, (AI) technology will bring economic benefits of tackling gender inequality, a new global competition, Chinese innovation and digital globalization.

Nowadays, advances in robotics artificial intelligence, and machine learning are in a new age of automation, as machines match or outperform human performance in a development to any countries. For example, automation of activities can enable businesses to improve performance by reducing errors and improving quality and speed, and in some cases achieving outcomes that go beyond human capabilities. For example, some research indicated automation could raise productivity growth globally by 0.8 to 1.4 % annually; more than 2,000 work activities across 800 occupations. When less than 5% of all occupations can be automated using demonstrated technologies about 60% of all occupations have at least 30% of constituent activities that could be automated. Many occupations will change that will be automated away: Activities most susceptible to automation involve physical activities, in highly structured and predictable environments, as well as the collection and processing of data. They are most prevalent in manufacturing , accommodation and food service and retail trade and include some middle-skill jobs. For example, such as natural language processing is a key factor. Beyond technical feasibility, the cost of technology competition with labor including skills and supply and demand dynamics, performance benefits including and beyond labor cost savings, and social and regulatory acceptance will be affected by (AI) automation technology. Thus, (AI) automation will impact to influence global employment in those aspects as below:

Firstly, assuming that people are displaced by automation will find other employment. The anticipated shift in the activities in the labor force is of a similar order as the long-term shift away from agriculture and decreases in manufacturing share of employment. Both of manufacturing and agriculture industries which would be accompanied by the creation of new types of work not foreseen at the time.

Secondly, for business, the performance benefits of automation are relatively clear. Thus, the businessmen have opportunities for their micro economies to benefits from the productivity growth potential and macro economies to benefit to encourage continued progress and innovation , investment and market incentives. At the same time, employers must innovate policies to help workers and institutions adapt to the impact on employment.

This will likely include rethinking education and training, income support and safety nets , as well as support for those dislocated, when employees need to leave themselves homes to move to other cities to learn new (AI) automation works. Thus, individuals in the workplace will need to engage move comprehensively with machines as part of their everyday activities, and acquire new skills that will be in demand in the new automation age. Consequently , the scale of shifts in the labor force over many decades that automation technologies can be a similar order to the long -term technology -enables shifts in the developed countries' workforces away from agriculture in the 21 th century. Those shifts did not result in long-term mass unemployment because they were accompanied by the creation of new types of work not foreseen at the time. However, human will still be needed in the workforce when the total productivity gains are caused by (AI) technology.

What occupations will be influenced by (AI) technology.

In the future, scientists predict that these occupations will be influenced by (AI) technology mostly. They include : retail salespeople, food and beverage service workers, language or translation teachers, health practitioners. Since these work activities have a more relevant occupations are made up of a range of activities with different potential for (AI) automation . For example, a retail salesperson will spend more time interacting with customers, stocking shelves , or ringing up sales. Each of these activities is distinct and requires different capabilities to perform successfully.

Thus, these job activities have similar simple control characteristics. Simple activities include greet customers, answer questions about products and services, clean and maintain work areas, demonstrate product feature process sales and transactions. All these activities can have similar simple activities in order to (AI) machines can be learn how to do these activities from (AI) technology . For example, the capability perception includes sensory perception, cognitive capabilities, such as retrieving automation, recognizing known patterns( supervised learning), logical reasoning problem solving.

Thus, (AI) machine is such human, which has feeling and emotion, such as social and emotional sensing, judgement reasoning methods, natural language understanding and physical capabilities, such as mobility , navigation, gross motor skill, fine motor skills. It seems that the future, (AI) human invents machines which will have these human characteristics to do human similar behavioral job duties more easily and efficiently. It implies these above human occupations will be replaced by (AI) human invention machines in the future. Due to (AI) creation, it is possible to cause unemployment number of these above workers will increase because (AI) machines can do their similar job behavioral activities.

Consequently, employers won't need to employ many of these skillful labor. Otherwise, they can buy less number (AI) machines to attempt to do whose job activities more easily and efficiently. So, it seems (AI) machines will have more high work performance to replace these occupation workers' work performance. Finally, these occupation worker unemployment number will only increase when the (AI) machines had been invented to achieve to do their work behavioral activities absolutely success in the future.

Whether (A) technology machine labor will replace human worker more or assist human worker more

There is no single agreed definition of a robot how outcome of a task that is completed without human intervention. When some definitions require the task to be completed by a physical machine moves and respond to its environment, other definitions use the term robot in connection with tasks completed by software , without physical embodiment.

However, to answer the question : Whether (AI) technology machine labor will replace human worker more or assist human worker more. I shall indicate some examples to let readers to judge whether (AI) technology can create new jobs or reduce old jobs.

Firstly, I shall explain what (AI) function is. (AI) is a service robot that performs useful tasks for humans or equipment excluding industrial automation application . Thus, the classification of a robot into industrial robot or service robot is done according to its intended application. It is also a personal service robot or a service robot for personal used for a non commercial task, usually by lay persons . Examples are domestic servant robot, and pet exercising robot. It is also a professional service robot or a service robot for professional used for a commercial task, usually operated by a properly trained operator. Examples, are cleaning robot for public places, delivery robot in offices or hospitals, fire-fighting robot, rehabilitation robot and surgery robot in hospitals. Thus, these functions will be future (AI) application to our daily life necessaries or business necessaries.

However, some authors agree (AI) will bring negative outcomes of automation, due to raise competiveness, reduce human job nature. Otherwise, other authors argue (AI) will bring positive outcomes of automation, due to raise productivities, job creation, assist humans work.

On the positive outcome hand, robots can increase productivity . This is particularly important for small-to medium sized businesses both are in developed and developing countries economies. It also enables large companies to increase their competitiveness through faster product development and delivery. Increased use of robot is also enabling companies in high cost countries to re shore, or bring back to their domestic base parts of the supply chain that will have previously outsourced to sources of cheaper labor. Currently , the greater threat to employment is not a automation, but an inability to remain competitive. Automation has led overall to an increase in labor demand and positive impact on wages. The reason is that the middle-income/middle-skilled jobs have reduced as a proportion of overall contribution to employment and earnings leading to fears of increasing income inequality, the skills range within the middle income bracket is large. Thus, robots are driving an increase in demand for workers at the higher -skilled and with a positive impact on wages. This issue is how to enable middle-income earners in the lower-income range to unskilled or retain. Finally, the (AI) positive impact supporter who argue the future will be robots and humans can work together.

However, on the negative outcome hand, robots can substitute labor activities, but don't replace jobs. They believe that less than 10% of jobs are fully automatable. Increasingly , robots are used to complement and augment labor activities, the net impact on jobs and the quality of work is positive. Automation can provide the opportunity for humans to focus on higher-skilled, higher-quality and higher-paid tasks. Robots can improve productivity when they are applied to tasks that which perform more efficiently and to a higher and more consistent level of quality than humans. For example, increased productivity is enabling some firms, such as Whirlpool, Caterpillar and Ford Motors company in the US restructure their supply chains, bringing back parts of the manufacturing process to the country of origin. Thus, productivity gains due to robotics and automation are important not just at the company level, but also for build industry and nation competitiveness.

I suppose that productivity can be raised. What are the impacts of robots on employment? Firstly, the main focus of development has been on personal entertainment, which does not drive worker productivity ( manufacturing production). When the internet ( information and communication technology (ICT)) innovation. This is borne and by findings that manufacturing productivity, which has been driven by innovations in automation rather than consumer technologies, has government strongly than productivity in the services sectors of the economy in most nature economies. It seems (AI) automation will create many jobs in internet communication entertainment game industry. For example, many young people like to use internet to play any electronic games from computer or mobile at home or outside home conveniently. Thus, (AI) automation will increase demand to be invented to any new entertainment game from internet channel. It will need to employ many (AI) entertainment game inventors to create many automation entertainment games. Thus, (AI) automation in internet entertainment game industry will need human (AI) entertainment game inventors to invent the knowledge-based capital of (AI) automation entertainment games. The (AI) entertainment game inventors will need own research and development skills, form specific skills, organizational know-how skills, databased knowledge, design and various forms of intellectual property to do these (AI) automation entertainment game invention occupations in the future.

International Federation Of Robotics(2016) indicated that China will be as a major robotics manufacturer and user of robots, benefiting from jobs created by robot manufacturing and productivity gains from robot use. Chins had sold of robots to any one single market every year since 2017 year. The Chinese government has included a focus on robotics in its 10 year strategy. In order to achieve its target of a robot density of 150 units per 10, 000 workers by 2020 year. Thus, Chinese companies will have to install around 650,000 new industrial robots between 2016 to 2020 year, 2.5 times more than installed globally in 2015 year.

Hence, China (AI) manufacturing industry will need to employ many workers . It implies (AI) manufacturing industry will create many new occupations in China. Also, ministry of economy, trade and industry (2015) also showed that Japan currently has the largest stock of industrial robots in operations, primarily in the automation industry. Driven by a rapidly aging population and low productivity rates, the Japanese government has sights on a 20-fold increase in the use of robots in the non-manufacturing sector and a three-fold growth rate of labor productivity in the service sector both by 2020 year. Thus, it also implies Japan will need many robots to be provide to service industry. Due to robots will provide to serve any businessmen's clients. Thus, it is possible that the service workers won't be dismissed as well as it is depended on the serving job nature to decide whether Japan's service workers can still serve to their employer when the service (AI) robots are applied to whose employers.

Consequently, it seems that (AI) can create employment, Ministry of economy, trade and industry (2015) showed that such as China will develop the major (AI) automation manufacturing industry. The (AI) employers will need to employ many workers to manufacture any these different kinds of (AI) robots to satisfy China or overseas individual or business buyers needs. But, (AI) can also cause unemployment to the low skillful service workers. Such as if Japan some service businesses choose to buy any (AI) service robots to replace their service staffs to serve their clients. It is possible that the service staffs will be dismissed, due to (AI) robots can do such as their same service job duties to achieve better service performance.

Thus, today, it is increasingly common for people to use robots in various situations at home and in retail stores, hotels and hospitals these service industries. Robots are classified into server types based on their functionality ( service and utility robots or those designed to communicate with humans) and appearance ( humanoid robots or mechanical robots). The type of robot, to which each country allocated particular importance in the advance of robotics, reflects the sense of values and preferences of its population. Thus, if the country has high population needs to use robots, then they will influence either more new jobs creation or more old job loss in the country's (AI) manufacturing or (AI) service industries both. For example, Japan respondents often associate the term " robot " with humanoid robots that can communicate with human and they have a high level of familiarity with robot. The US has the highest level of robot utilization at home and in retail stores with its people being the most enthusiastic about the future use of robots. Germany shows a strong tendency to consider robots for industrial purposes and its people feel strong effort to the presence of robots in their households.

In conclusion, to judge whether how (AI) will influence the country's employment to be better or worse. It will depend on the country home buyers (users) or business buyers (users) how to use (AI) for their daily needs. If the country , such as US retail stores need to use (AI) , it will have possible to reduce some or many retail service workers. Even, if the country , such as Japan has many home users need to use (AI) , it will not influence the employment market. Otherwise, it will raise (AI) salespeople numbers. Even, if the country, such as Germany and China will have many (AI) manufacturers, then it will create many (AI) manufacturing occupations for these (AI) manufactory workers.

Consequently, (AI) robots manufacturing and service needs will have positive or negative impact to any country's employment. It will depend on the (AI) service provision and service workers' job nature as well as the manufacturing workers of (AI) knowledge level to decide their employment chance in their country's employment market.

What does artificial intelligence(AI)
development stage mean

- What (AI) function is?

Some scientists explain that artificial intelligence means which is an expert system, computer software that embodies a portion of the specialized knowledge of a human portion in a specific, narrow domain, owns decision making ability of human expert. The (AI)technology is based on the premise that what makes a person an expert is years of experience that enables who recognizes certain patterns in a problem as being similar to pattern. For example, in the future artificial intelligence system can be applied to control air traffic, design to computer configuration, medical diagnosis, instruction/ training, speech/interpretation, monitoring to ( nuclear plant), planning to mission, factory scheduling, prediction weather, repairing telephone, automatic driving etc. different industries.

Artificial intelligence characteristics include: creative, adaptive , common sense, fact processing, quick replication, broad focus permanent and consistent skill. Otherwise, traditional computer expert system disadvantage includes perishable, unpredictable, slow reproduction, expensive, slow reproduction, slow processing lacks inspiration, needs instruction, narrow focus only machine knowledge. So, artificial intelligence is a branch of computer science devoted to creating computer to influence software and hardware to attempt to create human intelligence or human intelligent behavior. It is learning from experience, responds flexibility in situation that are, new or not anticipated.

Thus, (AI) can be learnt programmed knowledge to solve problems, using reasoning in solving problem, understanding and inferring facts and rules, recognizing the relative importance of different elements in a situation. In summary, artificial intelligence is concerned with two basic ideas mainly:

The first idea, it involves studying the thought processes of humans to understand what intelligence is; the second idea, it deals with representing thought processes using companies to create artificially intelligent entities for testing the theories of intelligence.

- Can (AI) impact human job nature?

Human need concern this question: Will artificial intelligence (AI) reduce some human jobs in order to instead of replacing machines to do? Due to artificial intelligence is the ability of machines to do thing, that people would require intelligence. For example, artificial intelligence machine man driving( self-driver), it (AI) machine man driving research is an attempt to discover and describe aspects of human intelligence that can be simulated by driving machine functions. Alternatively, (AI) mathematical research may be another viewed as an attempt to develop a mathematical theory function to describe the abilities and actions of things ( natural or man-made) exhibiting intelligent behavior and server as a design of intelligent calculation machine function.

Why do humans need artificial intelligence machines to instead of traditional human service job? For example, can artificial intelligence machine man (self-driving) driver drive to replace human driver? I shall compare the differences between humans and computers : The characteristics of humans are good at recognizing various things, either seen before or not, recognizing the relationship patterns between things. Human thinking is common sense reasoning, combining all types of sensory input, acting appropriately in novel situations, learning new things and changing behavior patterns, making decisions , even when given incomplete information, working with noisy, incomplete information gathering behaviors . However, characteristics of computers are good at: The tasks humans do naturally are extremely difficult for a computer program as intelligent, which must be able to do the same kind of tack as humans do naturally.

Hence, (AI) is an combination of many different success and technologies: Linguistics - computational and socio, philosophy-logic, philosophy of mind and of language, electronical engineering -image and speech processing, pattern recognition, robotics, machine learning, neural networks, optimization scheduling, management information system and decision making. So, it is possible that (AI) can impact human job nature to instead of human working behavior in the future.

- How can human society job nature to be changed to artificial intelligent society?

From the first intelligent perspective reason view point, artificial intelligence is making machines " intelligent" acting as humans expect people to act. Artificial intelligence has ability to distinguish computer responses from human responses, it owns knowledge to solve expert problem. From another research perspective reason view point, artificial intelligence is the study of how to make computers do things which, at the moment, people do better ( Rich & Knight, 1991, p.3).

(AI) researchers are native in a variety of domains, e.g. formal tasks ( mathematics, games), tasks ( perception, robotics, natural language, common sense reasoning), expert tasks ( financial analysis, medical diagnostics, engineering, scientific analysis and other areas).

From the second business perspective reason view point, (AI) is a set of many powerful tools, and methodologies for using those tools to solve business problems. From a programming perspective reason view point, (AI) includes the study of symbolic programming problem solving and search .

From the third human technological perspective reason view point, today's computer can do many well-defined tasks, for example, arithmetic operations, are much faster and more accurate than human beings. However, the computers' interaction with their environment is not very sophisticated yet. How can human test whether a computer has reached the general intelligence level of a human being? Can a computer convince a human interrogator that it is a human? But before thinking of such advanced kinds of machines, human will start developing our own extremely simple " intelligent" machines.

So, it is possible that human society job nature will to be changed to artificial intelligent society when (AI) technology is developed to the mature stage in the future.

- Why does human need artificial intelligence machines?

One of major division in (AI) is between humans who think (AI) is the only serious way of finding out how we ( human) work and human who want companies to do very smart things, independently of how we ( human) work. This is the important distinction between cognitive scientists vs engineers. One of another major division in (AI) is between symbolic (AI), which represents information through symbols and their relationships. Specific Algorithms are used to process these symbols to solve problems or deduce new knowledge and connectionist. So ( AI) , which represents information

in network. Biological processes underlying learning, task performance and problem solving are imitated from human mind behaviors.
Thus, it is possible that artificial intelligence machines can do the better judgicious behavior to compare human.

- How does artificial intelligence influence future working changing in automation employment and productivity aspects?

In the automation changing influence aspect, as companies increasingly use robots on production lines or algorithms to optimize their logistics manage inventory, any carry out other core business functions. Technological advances are creating a new automation age in which ever-smarter and more flexible machines will be deployed on an ever larger scale in the marketplace. However, researching artificial intelligence with how influences human working nature. We need to answer these questions: How will automation transform the workplace? What will the implications for employment? And what is likely to be its impact both on productivity in the global economy and on employment?
Advances in robotics, artificial intelligence, and machine learning are growing in a new age of automation as machines match or outperform human performance in a range of work activities, including ones requiring cognitive capabilities. What factors are determined the changing in workplace adoption by artificial intelligence innovation? What advantages are automation? Automation of activities can be enabled businesses to improve performance by reducing errors and improving quality and speed, and achieving outcomes that go beyond human capabilities.
Some scientists indicated based on their scenario modeling. They estimated automation could raise producing growth globally by 0.8 to 1.4 percent annually. Almost, the activities people are paid almost $16 trillion in wages to do in global economy have the potential to be automated by adopting currently demonstrated technology. According to their analysis of more than 2,000 work activities across 800 occupations. When less than 5% of all occupations have of least 30% of activities that could be automated. They also indicated that technical economic and social factors will determine automation. Continued technical progress, for example, in areas such as natural language processing is a key factor beyond technical feasibility , the cost of technology, competition with labor including skills, and supply and demand dynamics, performance benefits including and beyond labor cost savings and social and regulatory acceptance will affect ( alter) the scope of automation.

Other some scientists also indicate U.S. country for example, the anticipate shift in the activities in labor force of a similar order of magnitude as the long term sight away from agriculture and decreases in manufacturing. Share of employment in the United States both which were achieved. So, those factors can influence why artificial intelligence technology needs. So, it is possible that future agriculture and manufacturing both industries will apply (AI) technology manufacturer-kind of job nature to raise productivity instead of farmers, fruit picking workers, farming transportation labours as well as factory manufacturing workers and supervisors etc. human-kind of job nature.

- Is artificial intelligence possible to replace labor ?

Not just intelligence, but also debating, if machines are capable of having a conscious minds. Artificial intelligence has those characteristics as below:

On functionalism aspect, artificial intelligence inputs mental states, sensory inputs, ( beliefs, desires being in pain feeling) and behavioral outputs. Since mental states are identified by a functional role, which are thoughts to be manifested in various systems. Even, perhaps computers which are physical devices with electronic substrate that inform computations on inputs to give outputs similar to brains which are artificial intelligence composed of part any intrinsic relationship to each other. Thus, artificial intelligence activities is not the whole itself, but into parts or on external influence on the parts.

On dualism aspect, artificial intelligence is a set of views about the relationship between mind are matter. On materialism aspect, it builds the only thing that exists is matter, including consciousness.

On biological naturalism aspect, it is similar a human brain than feels pains makes mental situation. So, artificial intelligence is similar biologist which might to be excited to human labor work.

Hence, it seems artificial intelligence can change ( alter) or replace human labor work of nature in possible in the future.

- Can (AI) technology replace human labour nature of work?

On technological innovation reason view point, the history development of artificial intelligence studying the intelligence is one of most ancient scientific discipline. The history development of artificial intelligence what aims to achieve human use to sense, learn remember and think, logic probability, decision making and calculation develop from mathematics, instead of replacement human labor functions.

Artificial intelligence history development aim is the scientific analysis of skills in connection and practice with the appearance of computers from 1950 year beginning. The artificial intelligence (AI) can deal with the ultimate challenges. How can ( either biological or electronic) mind sense, understand and manipulate a world that is much simple and more complex than itself? And what if would human like to construct something with such capabilities?

The general-purpose software of the early period of (AI) were only able to solve simple tasks effectively and failed when which should be used in a wider range or an more difficult tasks. One of the sources of difficulty was that early software had very few or mix knowledge about the problems which handled, and activities successes by simply syntactic manipulation. Moreover, the other difficulty was that many problems that were tried to solve by the (AI) were untreatable.

The early (AI) software whether trying step sequences based on the basic facts about the problem that should be solved, experimented with different combinations till which found a solution. From the end the 1960 year, developing the so-called expert systems were emphasized. These systems had ( sue-based) knowledge base about the field which handled. Till to the beginning of the 1970 year, ( Prolog) the logical programming language was born, which was built in the computation realization of a version of the resolution calculus. ( Prolog) is a remarkably prevalent tool in developing expert systems ( on medical, judiciary and other scopes), but natural language parsers were implemented in this language. Then, in 1981 s, the Japanese announced the fifth generation computer system project, a 10 years plan to build an intelligent computer system that use the ( Prolog) language as a machine code. Nowadays, (AI) can be applied any industries, such as car manufacturing industry can use (AI) technological machine-men manufacture car, instead of replacing human labors in factory. Even, in the future, using (AI) machine-men drivers can drive any private cars or public transportation tools, instead of replacing human drivers, e.g. bus, train, tram, ferry etc. Also in the future, machine-men can replace housewives to serve families to do housekeeping clean job , e.g. cleaning toilets, bathrooms, kitchens, even cooking functions at home. So (AI) machine-man can reduce housewives works at home. Moreover, (AI) machine man can take care old people , when who are living at homes or elder care centers.

So, it seems artificial intelligence (AI) will be possible developed to manufacture a new generation machine-man to assist ( serve) families to do any simply cleaning or cooking jobs at homes. Moreover, the overall demand of ( AI) general social needs will also rise, such as security, driving transportation tools, restaurant cleaning, elder centers care service etc. So, it seems that individual or families or social needs of (AI) will be increase in the future. Thus, it will influence macro economy growth (GDP) if there are large house family consumer group and hotel or bus or taxis or ferry etc. different business consumer group demand any artificial intelligence machine numbers increasing. Then, the artificial intelligence products and material manufacturers must need to buy many artificaial intelligence materials to produce any kinds of artificial intelligence machines to prepare to satisfy consumer individual needs. Consequently, macro economy will grow to the owned artificial intelligence development countries, e.g. US, China, UK.

● Why can artificial intelligence satisfy human needs?

First, On machine-man satisfactory demand aspect view point, it makes computers that think, it is the automation of activities. We associate with human thinking: like decision making, learning. It is the act of creating machine that perform function that require intelligence when performed by people. It is the study of mental faculties through the use of computational models. It is the study of computations that make it possible to perceive, reason and act. It is a branch of computer science that is concerned with the automation of intelligent behavior. It is anything in computing service that human don't yet know how to do property.

Second, on thought aspect artificial intelligence means systems thank think like humans, systems that think rationally.

Third, on behavioral aspect, artificial intelligence systems that act like human and that systems act rationally. However, the basic objective of (AI) is to represent human's thought processes in computation . These machines are supposed to exhibit behavior that. It is performed by a human being, would be considered intelligent. However, some authors feel (AI) has disadvantages, such as it is not creative, it is excited in the use of sensory devices, it can't make use of a very wide context of experiences and it does not use common sense.

For speech recognition and understanding function needs example, (AI) can be applied in speech recognition and understanding function, which (AI) speech or voice recognition is a data input method. For example, the computer recognizes and understands one ( or a few) word commands. Speech understanding on the other hand is the computer's ability to

understanding a spoken language. That is , the computer understands the meaning of sentences, an paragraphs through (AI).
So, (AI) can be attempted to learn human language how to speak. It is similar to translate human language skill, instead of actual human speaking skill. Also, (AI) can assist handicap learning or language student how to listen different languages by machine-man sounds from computers more accurately.
So, it seems that it (AI) can replace human language teachers speaking function and can change teaching language nature of job in language speaking and listening education industry.

- Is artificial intelligence one good choice for human future technological benefit?

Nowadays, new technology development is popular. However, artificial intelligence is one kind of new technology choice among different technologies innovation. So it brings this question: Is artificial intelligence technology value to invest? To answer this question. I shall indicate some other new technology developments to compare (AI) technology development to judge which has urgent needs to achieve human expectation nowadays.
For example, why is green peace interested in new technologies? New technologies features prominently in our ongoing campaigns against genetic modified crops and number power. However, which are also an integral part of our solutions to environmental challenges, including renewable energy technologies, such as solar, wind and wave ( water) power energy as well as waste treatment technologies, such as mechanical, biological treatment.
It seems humans need concern how to apply (AI) technology to solve environment pollution challenges in our future. So, environment protective, agriculture, natural energy technology will be popular demand to attempt to apply (AI) technology to solve their challenges or apply (AI) to assist to develop their industry.

## How AI development stage to influence economy growth

How can artificial intelligence technology influence economy?
Advances in artificial intelligence (AI) technology and related fields have opened up new markets and new opportunities progress in critical areas, such as health, education, energy, economic development, social welfare and

the environment pollution.

(AI) automation will continue to create wealth and expand the global economy development in the future. However, when many will benefits that growth won't be costless and will be accompanied by changes in the skills, that workers need to increase productivity in the economy and structural changes in the economy. So, in the skills that workers need to succeed in the economy and structural changes.

I shall indicate why aggressive policy action will be needed to help Americans who are disadvantaged by these changes , due to (AI) technology is caused. For automation industry change example, artificial intelligence (AI) capabilities will enable automation of some tasks that have long required human labor. These artificial intelligence technology introduction can increase new opportunities for individuals. The economy and society, but (AI) has also the potential to disrupt be current livelihoods of many Americans. However, (AI) leads to unemployment and increase in inequality over the long run depends not only on the (AI) technology itself, but also on the institutions and policies that are changed.

Thus, it is possible that (AI) technology will raise some countries unemployment number if the employer apply (AI) technology workers to work instead of human labor in their factories, but it can also raise productivities for these employers.

- Can (AI) influence global economy growth?

Technological progress is main driver of growth of GDP per capita, allowing output to increase faster than labor and capital . However, technology can increase productivity, but also decrease the number of labor hours needed to create a unit of output. So (AI) causes unequal to labor wage decreases, even reduces the number of labor to manufacture, e.g. artificial intelligence technology of automation car manufacturing industry; clothing manufacturing industry; plane manufacturing etc. high technology of artificial intelligence manufacturing method. But (AI) should be potential environment benefit, although it raises unemployment ratio. Moreover, it can rise production , due to many skilled craft were replaced by the combination of machines and lower-skilled labor. The result of (AI) technology introduction , it causes output per hour risen when inequality declined, driving up average living standards, but the labor of some high-skill workers was no longer as valuable in the market. Otherwise, if (AI) technology is continue developed to be success. Some routine intensive occupations will be loss, which focused on predictable, e.g. easily programmable tasks, such as switchboard operators, filing clerks, travel agents, and assembly line workers would be particularly replaced by new (AI) technology. However, at the same time, (AI) technology development will bring these benefits: improvement in education ( training (AI) technology scientists) , due to (AI) manufacturing technology needs are raising to businesses and institutional changes, such as the reduction in unionization and raising in the minimum wage to the (AI) manufacturing technology skilled labor in factories.

Because (AI) technology is not a single technology, but rather a collection of technologies that are applied to specific tasks, the effects of (AI) will be felt unevenly though the economy. It will bring some tasks will be most easily automated than others , and some jobs will be affected more than others, both negatively and positively. Finally, new jobs are likely to be directly created in areas , such as the development and supervision of (AI) as well as indirectly created in a range areas though out the economy as higher incomes lead to expanded demand.

However, if (AI) technology could dominate global labor markets. If labor productivity increases, do not influence into wage increases, then the large economic gains brought about by (AI) technology could be increased wealth inequality, due to employers can reduce production cost, but workers ( labors) wages will not be increased, even will be decreased. Hence, it seems the (AI) technology will bring disadvantages to labor market to cause unemployment or reduce wages in possible, although it can reduce employer individual salary (wage) expenditure and it can raise productivity.

- How can artificial intelligence impact global economy growth?

Artificial intelligence (AI) technology is a branch of computer science that aims to create intelligent machines that work and react like humans. So, (AI) is a technology that appears to impact ( influence) human preference by learning, understanding complex contents, enhancing humans in executing both routine and non-routine tasks. In the future, (AI) technology that can be virtual personal assistant, as well as it may exist, such as robots with human-like processing capabilities.

How can (AI) technology impact global economy growth over the next 10 years? During this time period, (AI) technology is predicted to have wide-ranging applications including: Machine learning that automates analytical model building by using algorithms that allow machines to operate without human assistance.

In global education aspect, potential applications include predicting cause-

and-effect relationships from biological data, identifying new drugs, self-driving cars, and protecting against fraud, improved natural language processing that allows computers to continue to better analysis, understand and generate language to interface with humans using natural human languages. For example, transcribing notes dictated by physicians, automatically drafting articles and translating text and speech. So (AI) technology can be applied to education aspect to improve humans‘ knowledge level.

In visual art aspect, (AI) machine vision that allows computers to identify objects, scenes and activities in images. Current applications of (AI) machine vision include providing objective descriptions for the blind seeing( visual) needs.

We except the economic effects of (AI) technology to include both direct GDP growth from sectors that develop or manufacture. (AI) technology and indirect GDP growth through increased productivity in existing sectors that employ some form of (AI). If (AI) technology is an increasingly critical component of more products, it will become an integral part of many people's lives. Thus, (AI)'s ability to influence economic activity, rather than the economic or development status of the region. (AI) has the potential to impact income classes and to bring significant gains to both developed and developing countries. For example, (AI) has the potential to optimize good production around the world by analyzing agricultural regions and identifying what is necessary to improve crop yields.

In estimating the future economic effects by (AI) technology innovation, it is important to note that it is challenging to accurately predict which applications of (AI) will ultimately be commercially successful. In micro level economic influence, we need to apply methodologies to estimate the economic effects of investment in firms developing (AI) technology since investment levels in a technology are a telling sign of the future potential of that (AI) technology.

- How can (AI) influence GDP of high income countries in the next ten years?

How (AI)'s development may affect the global economy over the next ten years. In fact, (AI) technology has the potential to affect business across the global in a wide range of industries in ways only a number of technologies have done in the parts. For example, (AI) technology's expected to be a useful tool for enhancing human capabilities and in some instances replacing functions, such as driving a car, adoption of broadband internet, mobile telephone, industrial robotic automation have served to enhance human capabilities.

However, significant public debate has focused on projections of (AI) technology's effect on the labor force. However, large companies prefer to invest in (AI) technological industry. For example, face book's (AI) research lab., google machine intelligence lab. and micro soft machine learning and artificial intelligence research division are all making advances in (AI) technology and investing in the industry's top talent. Additionally, between 2010 year and 2015 year, nearly $5 billion in venture capital funding invested in firms across the global developing and employing (AI) technology ( Facebook (AI) Research).

- How can artificial intelligence impact on workplace?

Modern information technologies and the labor economy growth of machines is powered by artificial intelligence have already strongly influenced the world of work in the 21 ST century. Computers, algorithms and software simplify every tasks and it is impossible to image how most of our life could be managed without them. How can be the information economy characterized by exponential growth replaces the most production industry based on economy of scales? What will the future world of work look like and how long will it take to get? Will the future world of work be a world where humans spend less time earning their livelihood? Alternatively, are mass unemployment, mass poverty and social distortions also possible scenario for the future, where robots, artificial intelligence systems play an increasingly central role? These questions concern how artificial intelligence further development . Can influence labor economy growth on workplace ? When the labor market has widespread impact on intelligence property, information technology, product liability, competition and labor and employment laws.

How (AI) technology impacts on labor workplace.

The future influence any organizations how labor economies use of (AI) can be analyzed, such as deep machine learning is based on a set of model high level data. Unlike human workers, the machines are connected the whole time in workplace. If one machine makes a mistake, all autonomous systems will keep this in mind and will avoid the same mistake the next time.

Over the long run intelligent machines will win against every human expert. Production robots have been replacing employees because of the (AI) technology. They work more precisely than humans and cost loss. Creative solutions like 3D printers and the self learning ability of these production robots will replace human workers, the automatic data recording and data processing, traditional back office activities are no longer in demand. Autonomous software will collect necessary information and will send it to the employee who needs it. Additionally, dematerialization leads to the phenomenon that traditional physical products are becoming software. For example, CD or DVDs are being replaced by streaming services. The replacement of traditional event ticket, e-travel ticket service products or hard cash will be the next step, due to the possibility of payment by smartphone. So, (AI) technology will impact human's daily life consumption behaviors in the future. For another example, transportation tools, such as boats and ferries and private vehicles will use sensors and navigating without human input. Taxi and truck drivers will become obsolete, the stock store applies to stock managers and postal carriers of the delivery is distributed by (AI) machine delivery method.

What is the relationship between (AI) and (CRM)?

- Can (AI) technology impact on customer relationship management (CRM) ?

Nowadays , (AI) is a technology almost as old as the computer industry itself, it is similar with the advent of personal assistants function to businesses and personal promotion channel, such as ( Amazon's Alexa, Apple's Siri, Google's Assistant) image recognition ( face book), personalized recommendations ( Netflix , Amazon). Those innovations have been driven by a increase in processing power, lower cost hardware, and the exploding creation and availability of data. It seems, (AI) technology can impact global customer service management method.

How to forecast economic impact modeling to (AI) will affect global economy? Can human forecast business revenue growth and job creation ( or destruction) based on (AI) applied to customer relationship management (CRM) activities? In addition to the economic impact on (AI) or (CRM) which can include an estimate of the economic impact attributable to sales forces customer base. What can economic benefits be brought to (CRM) from (AI) technology?

Artificial intelligence(AI) comprises a set of technologies that use natural language processing, machine learning, knowledge graphs, and other tools to answer questions, discover insights and provide recommendations. Computer systems can use (AI) hypothesize and formulate possible answers based on available evidence can be trained through the ingestion of vast amounts of content, and automatically adapt and learn from (AI) self mistakes and failures.

So, any business organizations (customer service departments) can provide efficient and effective customer relationship management of excellent customer service quality if which applied (AI) technology system. The different type of (AI) systems include: (AI) system platforms, machine learning (AI) based data preparation and enrichment tools, machine vision/ image recognition, voice speech recognition, text analysis and natural language processing, bots , e.g. face book website and virtual digital assistance solutions, social media pattern analysis , sentiment analysis, advanced numerical analysis (e.g. IOT streaming , machine logs), supporting technologies, knowledge base dialog management, Q&A processing etc. different (AI) technology system customer relationship management (CRM) tools.

(AI) (CRM) of activity can include these categories, such as: corporate marketing, marketing operation, field marketing, customer support, digital commerce, customer analytics, customer influenced product or service design, product or service pricing, finance information, presentation, customer billing, inventory , logistics and fulfilment support, partner management etc. different CRM tools.

(AI) technology of CRM has been carrying on plan different stages to achieve CRM personal assistant tool for businesses. The stages are such as, in the beginning stage of (AI) projects in place, implement now, pilot phase next year in the final stage of (AI) customer relationship management tools are foreseeable future. So, this CRM technology has been improved to plan in different stages every year to prepare to achieve full capacity of CRM service quality for businesses to use in the future.

Hence, how to develop an estimate prediction of the economic impact (AI) technologies could have CRM activities, which depends on gathering macroeconomic information on business revenue and the basic marketing of business revenue and the basic markup of business expenses by major functions ( customer support, marketing and sales , production etc.)

An economic impact model that can gather data together and forecast the results how (AI) artificial intelligence technology brings (CRM) customer relationship management benefits to businesses, e.g. surveys investigation includes IT spending by sample countries, GDP and population estimates and forecasts, revenue per employee and ratios of IT spend to GDP. Surveys

( questionnaire questions) of forecast results are influenced by (AI) impact can include: results are projected from surveys and rely on estimates are made by respondents on the expected financial improvements in categories of (AI) –assisted customer relationship management activities. The forecast assumes that these estimates are correct; financial estimates are based on estimates of "first year" improvement from full (AI) implementation; forecasts are from planning to implement any artificial intelligence of customer relationship management (CRM) projects, the improvement forecast is of categories of activity , e.g. corporate marketing , digital commerce, and customer analytics. They are not estimates of ROI for the (AI) software. They rely on conservative estimates to which each of these entities might affect company revenue, expenses or productivity. They also rely on estimates of the penetration of software in customer relationship management activities . Net new jobs created are based on the ratio of new revenue to jobs required to support that revenue . They can assume that 50% of the net new revenue will support increases in labor and the rest will go for capital and other operating expenses that may replace jobs lost to automation.

In the future, some of the ways in micro economic benefits to any organizations. (AI) technology is expected to impact CRM activities include: Spending up sales cycles, improving lead generation and qualification solving customer support problems faster ( raising service quality), helping companies improve brand campaigns and recognition, lowering costs of support calls when increasing resolution rates, lowering the cost of recruiting employees and partners, increasing revenue from optimized product marketing, optimizing price, distribution logistics and preventing loss through fraud detection. So, micro economic benefits view point, it seems that (AI) CRM technology can raise any companies economic benefits for care term.

Artificial intelligence enables machines or the in-build software to behave like human beings which allows these decisions and act. The advent of (AI) is leading , talking, making decisions and act. The advent of (AI) is leading to new technologies advances and transforming the economic and employment opportunities for humans in a positive way. (AI) related technologies can facilitate our live. For example, industrial robotics, robotic medical assistants, smart games, financial forecasting software, big data analysis, algorithms in health and bioinformatics, pilotless cargo places, drone ambulances and general purpose and workplace robots and others.

(Disruptors technologies: Advances that will transform life, business and the global economy).
Artificial intelligence also known as computational intelligence is defined as " the human –like intelligence exhibited by machines or software. It is theorized that intelligence of humans can be described and intelligence machines or software can simulate it. These machines software can be reasonable , learn, perceive and process information, like human mind and thus facilitate human life. They can think and act for us. So, artificial intelligence is an interdisciplinary field of study including computer science, neuroscience, psychology, linguistics and philosophy.
However, (AI) research and developments have economically impacted many industries, such as robotics, telecommunications, computer applications , health, finance, heavy manufacturing, transportation, aviation, e-service and e-commerce, military , music and movie, toys and games entertainment etc. industries.
In fact, many ideas, systems and technologies have been developing in the world of (AI) technology. However, which are net called or considered (AI) products, rather which are mentioned with their specific names, such as smart graphics, machine learning, e-commerce etc. ( i.e. this is called (AI) effect).

- How can (AI) technology influence digital economy?

Nowadays, (AI) related industrial applications will replace most human power in fields, including call centers, customer services and air cargo transportation. (AI) technologies also help weather forecasting based on repeated rainfall pattern ( data) recognition, through robotics ( i.e. floor cleaning, moving lawns etc.) transporting people and products with unmanned vehicles, sending space unmanned smart shuttles, developing robotic arms, predicting market values in stock exchanges by internet, making homes safer, helping elderly and disabled using robotic servants etc.

Among the (AI) related technologies , there are a few that significance for the impact on society and especially on digital economy . (AI) is particularly influential in machine learning. Such as robotics, transportation, finance, health and bioinformatics, e-commerce , e-games, big online data gathering and internet-of-things. For example, machine e-learning is based in bioinformatics and robots that can learn new skills for better caregiving in healthcare. What is machine e-learning? Machines can e-learn from e-data gathering, coming up generalizations and making decisions to act in certain ways from internet.

There are important applications , such as e-machine perception, electronic online natural language learning processing, online search engines, online bioinformatics, online brain –computer interface, online game playing, online robot locomotion, online advertising, online computations finances, online health monitoring, online DNA classification and decision making, online in chemistry –cheminformatics . So, online machine learning can positively impact productivity and it can enhance information and analytical system from (AI) online channel.

What is robotics? Robotics is one of the most strongly influenced fields in (AI). For example, heavy manufacturing industries, robots and used and man power is replaced for effectiveness, precision, and accuracy, especially in respective or dangerous tasks, including welding, assembling , picking and placing .

So, robots can acquire new skills or adapt the changing dynamic environment. Also, artificial intelligence can be applied in developing transportation. For example, automated vehicles, driver assistance systems , safety systems, collision avoidance systems and public transportation. Moreover, (AI) technology has proven to produce some of the best tools to predict stock market fluctuations from internet data gathering method. It's predictions are based on ever-evolving predictions algorithms and systems learn new models and make connections between historical data and new data to measure stock market trading more accurate from internet data gathering channel.

In health field, especially in health data processing , analysis, decision making support and medical diagnosis. So, online data can show which patients will need what treatment and what alternative drugs could be used more accurate from (AI) online data gathering method. Bioinformatics is an interdisciplinary field combining statistics, (AI) online technology can help in discovering data patterns and modeling through the application of machine learning, artificial neural networks and genetic algorithms. For example, further (AI) technology development of human genome project of online data sequences.

Online shopping can be facilitated by virtual assistants developed through (AI) technology and these assistants can offer the best advice. (AI) online purchase coming after every product image recommendations and personalization bring important revenue to shopping online sites, like

Amazon . Smart computer graphics and games, artificial intelligence is useful in smarter computer, graphics, scene modeling , scene rendering processes in order to create, for example, effective human –robot interactions , online machine learning, online strategic games techniques etc. online computer related (AI) software.

So, online big data analysis and big data does have a critical need in the world of online intelligence machines and software in our future. In other words, (AI) offers online technology to enable online big data analysis to provide industrial organizations with valuable information for effective decision making in short time. For example, what IBM's Watson achieved: this machine used 200 million of structured and unstructured content with a special technology of hypothesis generation, massive evidence gathering, analysis and scoring from internet channel.

Finally, (AI) online technology another related internet invention ( internet of things) (IOT) is the network of machines or objects connected through internet. These connected objects can sense their internal and external environment, communicate with each other, can send critical data and finally can make decisions to act or correct their environment from (AI) online technology. For example, factories can monitor and automatically change production processes, hospitals can monitor and regulate the health conditions of their patients , schools can collect data from facilities and cars can send data to car makers from (AI) online technology.

Partner predicts that (IOT) market will create about trillion amount value by 2020 year. Although machines collect big data from their environment, whether which gain an insight or learn from these online data largely depends on the (AI) online machine learning principals and (AI) online technology. In 2013, Mckinsey estimated that disruptive technologies closely related with potential economic impact in 2025 year between $7.1 to $13.1 trillion amount ( automation of knowledge work, advanced robotics, autonomous or near-autonomous vehicles).

What is the relationship between
(AI) and global digital economy development ?

● Could work activities in China be automated
making in the nation with the world's largest automation potential?

Can (AI) technology influence China economy? Could China workers be affected and jobs made up of routine work activities and predictable? Will programmable tasks be particularly impact to China employment market

? When impact on labor market is likely to be gradual at the aggregate level, it can be sudden and dramatic at the level of specific work activities, rending some job obsolete fairly. Overall (AI) technology will raise digital skills when reducing demand for medium incomer inequality for China workers. It seems (AI) technology's effect on productivity could be crucial to China's future economic growth as the population ages are increasing.

In China, some biggest technological companies driving significant investments in research and development. Moreover, China is one of the leading global (AI) technology development county. However, China will need to focus on building its innovation capacity. For example, United States and United Kingdom are currently producing more influential (AI) technological research. However, if China planed to achieve (AI) technology success, it's traditional industries will need to develop technical know-how –to and overcoming implementation costs prepare to develop (AI) . When (AI) technology is introduced into China society, China government needs to raise concerning ethical, legal, technological security etc. business questions. Also, surrounding issues include privacy, discrimination, legal liability and regulation. It aims to encourage overseas investors to choose to invest (AI) technological industry to raise GDP growth and manufacturing industries income growth for long term in China.

If China encouraged overseas (AI) technology investment in its country. It is possible to influence China employment market to be changed. Because (AI) technology will impact to influence China people daily life. Due to (AI) technology is introduced to China society, many rich people will prefer to spend to buy any high (AI) technological products for entertainment or learning or machine man driving etc. daily necessity activities. Then it will raise GDP growth and will raise (AI) manufacturers or related-(AI) technological manufacturers profit. It is beneficial to China because it can become one high knowledgeable and (AI) technological economical society.

But it will bring bad influences to raise unemployment chance for the low skillful labor. In labor economy aspect influence , how (AI) technology can influence China low skillful labor unemployment ratio raising. The raising low skill labor unemployment reason is because China low skillful human labors are argued or are replaced by (AI) technology creating new challenges to introduce to influence China society of simply human manufacturing job nature to be changed to be high (AI) technology manufacturing job nature in any China factories. Moreover, when (AI) technology introduction to China, it will cause other related social challenges in China. The varied

(AI) related challenges, including the difficulty of creating safe and reliable hardware for sensing and affecting ( transportation and education), the challenges of gaining public trust, a low resource comities and public safety and security, the challenges of overcoming fears or marginalizing humans in China employment and workplace and the risk of diminishing interpersonal trust because the low skillful labors won't believe any China employers will give chance to employ them , due to (AI) technology will replace their skills and man manufacturing of productivity is much less to compare to (AI) technology manufacturing method.

- How does ( AI) technology influence
the future of employment change?

Are future nature of jobs changed to computerization from (AI) technology? Where are the probability of computing occupations from (AI) technology influence? What is expected impacts of future computing on labor market from (AI) technology influence? John Maynard Keynes's frequently cited prediction of widespread technological unemployment " du to our discovery of means of economic the use of labor outrunning the pace of which we can find new used of labor" ( Keynes, 1933, p.3).

In the future, (AI) technology will impact some nature of occupations to change computing. This chance will also influence some countries' economic change. For example, some factory human labors hand routine manufacturing tasks will be changed to computerization of routine manufacturing tasks by (AI) technological machine men hand manufacturing method. it will cause a structured shift in the labor market, with workers reallocating their labor supply from middle-income manufacturing to low-income service occupations.

Arguably, this is because the manual tasks of service occupations are less computerization, as who require a higher degree of flexibility and physical adaptability. So, (AI) technology will influence the human hand labor skillful occupation nature of task cheaper , such as vehicle manufacturing , ship manufacturing, computer manufacturing, steel manufacturing, television, radio etc. home electronic products of heavy machine industry change. Due to (AI) technology machine man will be proper to be used to manufacturing these electronic products when the (AI) technology innovation can develop to the mature stage. Then, any countries manufacturers will choose to use (AI) technology machine man, instead of human hand production.

Supposing the future prices of computing are fallen, seriously, problem solving skills are becoming relatively productive, explaining the substantial employment growth in manufacturing occupations, involving cognitive tasks where skilled labor has a comparative advantage, as well as the increase education needs for (AI) technology computing of machine man subject study.

Prediction of education needs for (AI) technology student numbers will increase, due to manufacturing industry needs many (AI) technology students in future employment market. Another (AI) technology influence if the future (AI) technological innovation, e.g. machine man manufacturing or machine man service industries will both increase demand, then with more sophistic software technologies will be disrupted labor markets by marketing workers redundant.

For publishing industry, what is striking about the case in paper book publishing industry will be unpopular? Due to the electronic book publishing industry will be popular, e.g. Amazon publish . (AI) technology can influence paper book manufacturing method which is replaced by machine man electronic book manufacturing method as well as it will cause the computerization is no longer confined to routine manufacturing tasks. Due to (AI) machine man manufacturing technology will be proper to be used to manufacture any products in short time efficiently and effectively , e.g. electronic book products. In the future, if it is fact to occur this case, such as ( AI) technological machine man manufacturing method will be adopted ( applied) to manufacture electronic books or any products in possible. (AI) technology will cause many manufacturing workers are unemployed. It is beneficial to employers, who can reduce to spend much wages expenditure to employ manufacturing workers, but it will cause many manufacturing workers loss jobs and reduce income to support whose families lives. It will cause social challenges, e.g. increasing stealing crimes if the manufacturing workers had not other skills to find other jobs to do easily. So, manufacturers need to concern over technological unemployment which will be hardly future phenomenon if who decided to dismiss all manufacturing workers, due to (AI) technology machine men replace to them.

If ( AI) technology can be innovated to produce any kinds of machine man to serve any service or manufacturing industries successfully. Then, it will bring these questions: Can future that workers be influenced to be automation employment and productivity by (AI) technology influence? Does it impact to influence the (AI) technology countries' productivity and growth and natural resources development and labor markets and evolution of global financial markets and economic impact of technology and innovation and urbanization etc. issues? How will automation transform the workplace? What will be the implication for employment? What is likely to be its impact both on productivity in the global economy and on employment?

In fact, automatic of activities can enable businesses to improve performance by reducing errors chance and improving quality and speed, and same cases achieving outcomes that go beyond human capabilities. Some economists indicate (AI) technology would give a needed boost to economic growth and prosperity have of the working age population in many countries. Based on the scenario modeling, they estimate automation could raise productivity growth globally by 0.8 to 1.4 % annually. They also indicated that almost half the activities people are almost $1.6 trillion in wages to do in the global economy have the potential to be automated adapting current demonstrates technology, according to their analysis of more than 2,000 work activities across 800 occupations. When less than 5% of all occupations can be automated entirely using demonstrated technology, about 60% of all occupations have at least 30% of worker made activities, that would be automated. More occupation will change to be automated. They also indicated for business performance benefits of automation are relatively clear, but the issues are more complicated by policy making to attract foreign investors. Beyond technical feasibility, the cost of technology, competition labor will include skills and supply and demand dynamics, performance benefits and beyond labor cost savings and social and regulatory acceptance will affect the automation. Their predictions suggest that half of today work activities could be automated by 2055 year, but this could happen 10 to 20 years earlier or latter depending on the various factors in addition to their wider economic condition.

Some scientists suggest (AI) technology is finally starting to deliver real-life business benefits. Computer power is growing significantly , algorithms are becoming more sophisticated and perhaps most important of all, the world is generating vast quantities of the fuel that powers (AI) technology data billions of gigabytes of it every day. Also, online firms are digital natives, such as Google online search service company is investing on (AI) technology. For new though most of the news if coming from the suppliers of (AI) technologies. And many new users are only in the experimental phase. Few products are on the market or are likely to arrive these soon to drive immediate and widespread adoption. As a result, analysts believe

(AI) technology's potential will give true economic benefit in the future. (AI) industry will introduce to suppliers and users to raise economic potential of (AI) technology.
In the future, (AI) technology systems can solve business problems. Some scientists categorized those into five technology systems that are key areas of (AI) technology development: robotics and autonomous vehicles, computer vision language virtual agents and machine learning , which is based on algorithms that learn from data without replying on rules-based programming in order to draw conclusions or direct an action.
Such as computer vision and language includes natural language processing, analytics, speech recognition technology, some are about learning from information, such as about machine learning and others are related to acting on information, such as robotics, autonomous vehicles and virtual agents, which are computer programs that can converse with humans. Machine learning and a subfield called deep learning are artificial intelligence applications.

- Can artificial intelligence impact

global economy growth?
Artificial intelligence (AI) is a term first defined in 1956 year. It is a branch of computer science that aims to create intelligent machines that work and react like humans. In contrast today, 60 years later, (AI) is characterized by a number of applications, including computers playing games against humans and understanding human languages, virtual personal assistants, and robotics which involve computers seeing , hearing and reacting to sensory stimuli. In the future, technologists predict for (AI) technology ranging from (AI) being used as a tool to aid relatively simple processes for robots with human like mental capabilities, who expect (AI) technology can emulate human performance by learning, coming to mind its own conclusions, understanding complex content, engaging in dialog with people, enhancing human cognitive performance or replacing humans in executing both routine and non-routine tasks. In existing industry, (AI) technology is used , such as targeted advertising and virtual used personal assistant as well as the (AI) technology that my exist in the future, such as robots with human vehicle processing capabilities.
The range of (AI) technology's progress in the future will determine the economic impact future of (AI) technology on the global economy with more limited advances and applications ( i.e. weak (AI) only) corresponding to more limited economic impacts and more substantial progress, i.e. strong

(AI) technology is corresponding to more significant economic impact.

(AI) technology learning that automates analytical model, including predicting cause-and-effect relationship from biological data, identifying new drugs, self-driving cars and protecting against fraud etc. functions. Also (AI) learning can improve natural language processing that allows computers to continue to better analyze, understand and generate language to interface with human using the natural human language, virtual personal assistant, helps users by providing scheduling appointment, reminds organizing personal finance and finding providers of various services, machine vision allows (AI) machine man to identify object, scenes and activities in detect pedestrians and bicyclists.

We expect the economic effects of (AI) technology to include both direct GDP growth from sectors that develop or manufacture (AI) technology and indirect GDP growth through increased productivity in existing sectors that employ some from of (AI) technology. If (AI) producing sectors could grow, then it could lead to increase revenues and employment of (AI) technological professionals within these existing firms as well as the potential creation of entirely new economic activities to any countries' societies productivity improvement in existing sectors could be realized through faster and move efficient processes and decision making as well as increased (AI) technological knowledge and access to information available in societies easily.

In the future, if (AI) technology is an increasingly critical component of more products, it will become an integral part of necessary products of many people's lives. The extent of (AI)'s economy effort is also likely to vary from region to region, thought variation may be more dependent on the predominate economic activity of a region and the (AI) ability can influence economic activity, rather then the economic or developmental status of the regions. (AI) technology can move accessibility and can use source development to do international business between one country and another country.

So (AI) technology has the potential to give benefits to different income chooses and to bring significant gains to both developed and developing countries. For agricultural technology, (AI) has the potential to optimize food production around the world by analyzing agricultural regions and identifying what is necessary to improve crop yield. In total, (AI) technology gives greater economic impact to any countries agricultural regions if which implemented (AI) technology to grow crop , fruit etc. food production in the farms.

Investment in (AI) technology is such as capital investment to any countries' public or private enterprises. So, it will have large economic impact to the future . If the (AI) technology is reasonable invested to the different needs aspect by the public or private enterprises in the country. Then, it will have good economic impact to the country in the future. However, when (AI) technology is likely to affect both the productivity and employment components of economic growth in many sectors. Significant public debate has focused on projections of (AI)'s effect on the labor force. However, for instance, some researchers have argued that the rise of (AI) technology and automation will led to significant unemployment as capital is substituted for the low skillful labor. So, they point to the concern that the increasing sophistication of (AI) technology may balance skilled and semi-skilled workers and the reduce the size of the middle class. However, this is not a new argument, due to (AI) technology negatively affecting the labor force and leading to mass unemployment. Because the (AI) technology is the substitution of machinery for human labor. Although, employment in certain industries, has been reduced in the past due to technological advancement. For long term, the labor market has adapted to the introduction of new technology, giving rise to new jobs in new areas. (AI) technology may also be accomplished without a reduction to total employment in the long-term to some Asia countries, such as Hong Kong and Japan. Because Hong Kong and Japan many low skilled labor, e.g. security, cleaner who complaint that employers need them to work long time hours. ( abnormal working hours) e.g. one day 12 to 15 working hour per day. Hence, if (AI) machine means invention technology success. Security or cleaning job can be worked from (AI) machine man in some hours every day in order to reduce the long time working hours cleaners or security workers, e.g. one (AI) machine man works 4 hours for cleaning or security job, one day as well as another cleaner or security labor only needs to work 8 hours one day. So total security or cleaning employers can employ 12 hours machine cleaners or security workers and human cleaners or security workers in one day. For long term benefit, Hong Kong or Japan every security or cleaning worker does not need to work 12 hours minimum working hours one day. They won't feel tried and bore and without private with whose families, so who will accept to do these cleaning or security jobs, even they can raise work efficient and performance when who feel happy and health. So, (AI) technology of machine man invention can raise low skillful labor efficiency and it can help them to avoid abnormal working hours demand in some busy work life countries, such as Hong Kong and Japan. Before, one Japan female labor feel unhappy to work, due to who often needs to work abnormal working hours for her employer and who has less sleeping and without any private time to enjoy her life with her families every day. So this abnormal working hours factor causes her to do commit suicide behavior, then she is die unlucky. So (AI) technology of machine man invention ought avoid abnormal working hours demand for employer in any countries in the future.

The most important occurrence to any employers, some researchers had attempted to do one experiment to find that private research and development , venture capital and public research and development investment all have strong net effect or economic growth with venture capital funding further having the strongest such effect from (AI) technology. The researchers hypothesize the venture capital investment contributes to economic growth through (AI) technology innovation and by the capacity of an economy to use existing (AI) technology knowledge to increase productivity. They predict the impacts of venture capital, business-research and development and public research and development can raise multi factor productivity from (AI) technology introduction.

Can (AI) technology influence the economic development to developing countries? The developing regions of the world contain most of natural resources. If one day, (AI) technology has invent one kind of machine man which can assist any gas or oil workers to seek any new oil/gas natural resource locations easily. I believe that (AI) technology can help these natural resource exploitation countries will gain economic benefit more easily. So, (AI) driven technology can be used to change to create any new opportunities to address poor management or resources and improve human well being, such as Africa Latin America and India can use (AI) technology machine man to seek any oil/gas natural resource countries exploitation activities to attempt to gain much economic benefits.

● Why will (AI) technology grow economic
development ?

Nowadays, increases in capital and labor are no longer driving the levels of economic growth, such as (AI) technology. The ability of increase in capital investment and in labor of traditional drivers of production, have no longer to be enjoyed in most developed economies ,e.g. developed country, US, UK

. However, artificial intelligence has the potential to overcome the physical limitation of capital and labor to avoid missing out on this opportunity. So, policy makers and business leaders must prepare for and work toward a future with artificial intelligence. They must do with the idea that (AI) is another simply method to enhance productivity method . Rather they must see (AI) as the tool that can transform thinking about how growth is created.

Economists have always thought of new technologies are as driving growth their ability to enhancing. It can replace labor and capital factor of production. So, it brings this question: What is the factor of production (AI) technology characteristics. They key factor is to see (AI) technology as a capital-labor .
(AI) can replicate labor activities at much greater scale and speed, and to even perform some tasks began the capabilities of human. For example, by using virtual assistants , 1000 legal documents can be reviewed in a matter of days instead of taking three people six moths to complete. Some (AI) technology may be one kind of factor of production in the future. For another example, people will work in workplace digitalization environment. So, in the future, working environment and information management are automated. Such as Konica camera sale company will use workplace digitalization. So , (AI) technology can provide workplace digitalization in order to raise productivity efficiency. (AI) technology will be one kind of production which is replaced by workplace digitalization and it will grow any organization productivity efficiently. Then, (AI) technology will assist overall social economy growth , due to productivity is raised and products can be produced in short time to prepare to sell in consumption market. So, time will be shortened to increase GDP growth fast for the development of (AI) technology countries.

● How can (AI) technology impact to global economic and social and psychological changes?
What will be the development of (AI) technology and predictions concerning the future evolution? The computers and robots will develop conscious, intelligent and minds into humans, enhancing psychological and behavioral abilities and allowing for direct communication with (AI) minds. (AI) technology will be impacted human life by (AI) technology information communicative and environmental influence. A " world brain" and " world

mind", this psychological system will be enhanced and enriched the capacities of both individual and collective cognition by (AI) technology of service industries.

(AI) technology with influence these human needs of service industries changes, such as , biological science, finance, entertainment, business, biological science, transportation, communication military etc. The personal computer evolution, the internet and the world wide web which exploded on the scene, linking business, homes, schools, social organizations which were a completely unpredicted phenomenon to influence human life. Kurzweil (1999) predicts that by 2029 year, most human communication will be with machines. According to Person, by 2100 year, there will be human machine convergence.

How can (AI) technology influence environmental protection to make benefits to farming economic growth? (AI) technology can be applied to predict how to solve environmental pollution challenge to avoid to damage any crop or vegetable or rice or fruit etc. food growth. Because environmental experts can gather global environmental pollution data from an environmental database to build a perform a systematic analysis from (AI) technology. The first step is this broad analysis can include understanding, statistical and data gathering techniques to obtain the relevant data, the correlation among the variables involved, and a list of possible models. The next step is to select a set of methods and models that cover all kinds of knowledge and functionalities needed for the decision making process. Once the models are selected, they must be fully implemented by means of machine learning , data mining, statistical or numerical technique. After that, those models must be integrated to build the whole EDSS. The EDSS must be tested to check its performance, accuracy, usefulness and reliability, both from the user's and (AI) technology/computer scientist's point of view. If these is any wrong feature in any development stage, such as model's integration, models' implementation, selection of models, database, problem analysis etc. the developers must come back in the update th required components. When the evaluation phase is all right, the EDSS is ready to be applied to the environment. The great contribution of artificial intelligence to EDSS the integration of several methods complementing the classical statistical models/simulation , statistical analysis, linear models, etc. and numerical models ( control algorithms, optimization techniques etc.) .

This cooperation makes the resulting systems more reliable and powerful in coping with real world environment systems. Date interpretation has been a principal area of research in (AI) technology since the very beginning. The most demanding problem in the environmental assessment context. Knowledge representation permits the definition of the different types of data that the existing methods adapt to the process. There is also a lot of work to clean, repair and transform the huge available quantities of raw data. Apart from this, the availability of meta-information or background knowledge is required to guide the process. Data mining is multi-disciplinary: It covers expert systems, data based technology, statistics, data visualization and unsupervised machine learning. These techniques operate at the level of data and background information, where numerous and often incompatible new commensurate pieces of information from disparate sources have to be brought together ( K, Fedra, 1994).

So, it seems that in the future, (AI) technology with the increasing maturity in particular those related to knowledge and engineering, new dimensions can be assisted to users in environmental decision making are available. For example, many environmental systems are characterized both by incomplete models and by limited data. Hence, in the future, (AI) technology will be applied to predict climate change to reduce crop or fruit etc. food agriculture challenge by climate change bad influence.

- Will (AI) technology influence digital economy change to manufacturing industry ?

To understand how the manufacturing business must adapt to prosper in the technology, we need to understand how (AI) technology will change us to shape our daily habits to satisfy our expectation of products to how we shop and even the immediate of the entire process. For example, taxi services are in the crosshairs as on demand transportation services like, available of the touch of a smart phone button expand. In fact, Yellow lab, US country , san Francisco city's largest taxi company is filing for bankruptcy as the industry starts to change faster than almost anyone expected. However, at this point, its more than an app that is changing, some our taxi passengers renting taxi transportation to catch consumption behavior.

(AI) technology will influence digital economy for taxi passenger's individual customer experience, offering a growing renting taxi to catch of service and feedback opportunities when any one taxi passenger who chooses to use mobile phone app online tool to prepaid to rent any taxi more easily.

Also in the long term, (AI) technology can influence vehicles drive themselves of behavior. Already, companies like Google and GM are working on projects to bring fleets of autonomous vehicles to cities at the path of a button.
Moreover, this on-demand service model is beginning to appear across a much broader range of markets. For example , Amazon company is investing in its own fleet of trucks, planes and even drone at the same time as it pushes for same-day delivery of products. As some point, vehicles will be autonomous too. So, it seems that (AI) technique will influence any transportations choose to use digital autonomous driving technology in the future . For Amazon company case, it is not stopping of logistics. It is also aiming to automatically manage the supply of consumer home products with its recently launched Amazon replenishment service, Dash. Dash is a digital service that enables that connected derive to automatically order physical products from Amazon when supplies are running low. So, it seems (AI) technology will be applied to logistic function by digital technology method introduction in the future.

Hence autonomous vehicles will optimize industry supply chains and logistics operations through increased efficiency and flexibility. In fact, fully automated and lean supply chains will keep reduce load sizes and inventory by leveraging smart distribution technologies and smaller autonomous vehicles by machine man assistance. If Amazon continues to grow market share for online sales by reducing effort required by the consumer to place an order, when also contributing the almost immediate delivery of products to the doorstep. So, it will further fuel the trend toward on-demand derive. As Amazon company fuels the on-demand economy, consumers will expect immediacy in more parts of the digital economy. On top of speed, consumers increasing expect more personalization options.
So, (AI) technology will influence digital manufacturing, such as Amazon publishing to monitor every aspect of every process in real -time and communicating to self-optimized deep learning robotics, new methods of high volume and high customization will become possible. Then, as products merge into product platforms and even services, manufacturers have the opportunity to provide components and platforms used by smaller players. So, (AI) technology will influence manufacturing industry to choose automated SMI lines, robots installed, automation engineers.

Another future (AI) technology development can be applied to space science aspect, such as Automation engineering space in manufacturing process to achieve digital manufacturing benefits to any businesses in the future. Such as reducing cost, shortening manufacturing time, raising efficiency, shortening delivery products to client individual time. How can artificial intelligence give the need and advanced fast and evaluation methods benefits for space exploration? When US NASA ( space exploration organization) achieves any space exploration missions, it will answer this question:

When is it useful to have a machine use (AI) technology to achieve a decision? After all, after millions of years of space exploration and rough 10,000 years of civilization, humans are usually quite good at making decisions in complex uncertain environments. Through, Johns Hoplains University's Applied Physical Lab. Research in (AI) technology enabled systems, which has identified three general use cases for (AI) technology to explore space mission:

First, for some tasks (AI) technology is more cost effectiveness than human. Second, (AI) technology is better suited than humans at solving some, but not all problems. Third, (AI) technology allows NASA organization's space exploration mission to develop machines that ate capable of responding faster than when a human is in the decision loop ( D. Scheidt, 2012, A. Castano et. al. 2008).

So, the use of (AI) technology to enable science by observing the pace of rapidly evolving phenomena was demonstrated. It is more effectively coordinating and (AI) technology utilizing to earn economic benefits to use for space exploration mission.

However, (AI) technology also have current risk for space exploration. Today (AI) technology is immature and requires further development to reach its potential. For instance, the (AI) technology algorithms that detected the dust derive could not have identified whether the Martain weather represented a threat to the cover. Also it can not yet use instrument input to determine what, where and how to autonomously make the next space science measurement. An equally important factor limiting (AI)'s deployment is that lacks the methodology and technology to effectively test (AI) technology. So, the challenge will testing (AI) enabled system is how (AI) performance can be measured. It would be NASA organization's difficulty to find (AI) technology to develop to carry on researching any space exploration missions in the future. However, (AI) technology will be a good economic benefit choice for space exploration mission in the future.

- What is artificial intelligence potential benefits and ethical considerations?

The ability of (AI) technology systems to transform vast amounts of complex information into insight has the potential to help solve manufacturing or service challenges for human needs. However, to reap the societal benefits of (AI) systems, humans will need to trust then and make sure that which follow the same ethical principles, moral values, professional codes and social norms that we humans would follow in the same scenario, research and educational efforts as well as carefully designed regulation in order to achieve the most effort of economic benefits goals. For example, international business machines corporation (IBM) is actively engaged both competitors , in global discussions about how to make (AI) ethical and as beneficial as possible for people as social economic benefits.

(AI) is usually defined as the " capability of a computer program to perform tasks or reasoning processes " that human usually associate to intelligence in a human being. Often, it has to do with the ability to make a good decision, even when there is uncertainty, too much information to handle. As an example, play chess or complex card games of entertainment activities is believed to need some form of intelligence in a human being, as well as choosing the best medical facilities in a difficult medical case, or creating something new, such as mathematical theorem or even some form of act, or even driving automatic machine man ( self driving vehicle) replacing human driving in the middle of a crowded city.

(AI) needs depends on what we consider being intelligence in the behavior of a human being act a certain point in time. If human belief about human intelligence changes and we don't believe any longer that a certain task requires intelligence, then a computer program performing that task is no longer part of (AI), it becomes just another boring computer program. So, it means that (AI) technology will replace some old computer programs, if human can invent new generation of (AI) software for any functions or activities to satisfy human needs.

As IBM, it argues intelligence. This means that we aim to build systems that enhance and scale human expertise and skills rather than replacing them. We therefore focus on practical applications of (AI) capabilities that assist people in performing well-defined tasks of needs by exploiting and wide range of (AI)-based services. We also use the term " cognitive computing" it is mean a comprehensive net of capabilities based on technology. It comprises the fields of machine learning, reasoning and decision technologies, language, speech and vision recognition and processing technologies, high performance and high efficient functions for any industries or individual consumers needs. For example, robotics, which are usually very good at doing what which are supposed to in any environment, much have public shopping center, factory etc. places which need simply services from the robot ( machine man), such as cleans the floor of our houses to the robot that can work together with humans in production chains, passing through the warehouse, robots can take care of the tasks of an entire warehouse and the companion robots like Nao, Pepper, Aibo and Giraff, who can entertain use, talk to use and help elderly people to stay connected to their friends, relatives and doctors.

Google company is building automatic machine ( self-driving cars ) and has acquired more than 10 robotics companies. Facebook had opened whole new research facility only on (AI) research. Apply computer has developed Siri. Microsoft computer company has built a similar personalized assistant. Google has Deep mind, a UK company whose long term aim is to build general (AI) and has already great potential to win game to the world champion and IBM is investing a huge amount of resources in applying its Watson cognitive computing system to the medical domains to finance and to personalized education. In Europe, IBM is establishing new centers in Munich and Milan focused in the application of cognitive computer capabilities to the internet of things and healthcare respectively.

For example, automatic machine man ( self-driving cars) are all about (AI), which used to be able to see what happens in the street ( signals ,lanes, other cars, pedestrians, traffic lights, which need to able predict what other cars and pedestrians will do, and who need to be able to cope with unforeseen situations. Since, most car accidents are due to human fault, it is estimated that the adoption of self-driving cars will save about half of the lives that are usually last in car accidents.

IBM Watson company has to understand spoken language, make sense of massive amount to text , respond correctly to questions in many categories, as well as assess its own confidence in responding to such questions. In the future, (AI) technology can own question/answering capabilities that would be very useful, for example, in assisting a doctor when trying to some to the correct diagnosis for a patient and to propose the best therapy .

Intelligent machines can also rely on huge amounts of data to be used to learn how to make better decisions. This data comes from all of us over the years Facebook users have uploaded more than 250 billion pictures and every day who upload about 350 million more. Every second, we submit 40,000 google search queries. So, (AI) technology will be connected through the web from appliances to traffic lights from cars to watches. Other tasks that are very easy for humans are physical and manipulation tasks, such as walking , running, picking up an object to make its shape and location, restricted environment. But (AI) machine man technology still not able to have the general physical and manipulation capabilities even of a 6 year old.

So, it brings this question: Why do (AI) scientists need to concern ethics? Because (AI) technology is complex, information into insight has the potential to reveal long held secrets and help solve some of the world's most difficult problems. (AI) systems can potentially be used to help discover insights to treat disease, predict the whether, and manage the global economy. So, ethic issues is important to and (AI) scientists . If any one new (AI) technology research investigation could success, it will be a secret to and the (AI) scientists can not permit to their loyalty to any competitors to damage the fair (AI) technology products trading market. The country ( countries) (AI) technology scientists need to concern ethic issues, who need to keep secrets for their countries economic or/and social benefits. This is moral issues to any countries/country loyalty is whose countries intangible assets. They can not sell (AI) loyalty to any their countries to assist whose economic benefits immorally.

- How can (AI) technology influence to global
health care economy development?

According to (AI) lecturer analysis, when combined key clinical health (AI) application can potentially create $150 billion in annual savings for the US healthcare economy by 2026 year. (AI) technology is re-winning modern conception of healthcare delivery. It enables machines to sense, comprehend, act and learn. So which can perform administrative and clinical healthcare functions (Accenture, 2017).

It will help health care service organizations to reduce health care cost, will improve and raise service quality and access. So, (AI) health market size will be predicted growth. (AI) applications in health care include robot-assisted surgery, virtual nursing assistant, administrative workflow assistant, fraud detection, error reduction connected machines, clinical trial participant identifier, preliminary diagnosis, automated image diagnosis and cybersecurity.

What kind of benefits (AI) technology can contribute to healthcare service? (AI) technology can deliver what many health care organizations need, such as financial and operational of labor costs, digital expectations from patient consumers how to use (AI) technology to solve interoperability challenges in any healthcare organizations. Also (AI) technology can be applied to wellness an d lifestyle management, diagnostics, delivers financially but also way of organizational and workflow improvement. So, (AI) technology will be continue to become most prevalent and adoption to healthcare organizations , which must need to enhance structure to be position to take full advantages of new (AI) technological capabilities. (AI) technology can change the nature of work and employment is rapidly changing to make the best use of both humans and (AI) talent in healthcare industry in the future. For example, (AI) technology offers a way to fill in gaps and the rising labor shortage in healthcare. According to Accenture analysis, the physicians shortage is increasing. However, (AI) technology will manufacture healthcare machine men to replace physicians in future one day(2017). Hence, (AI) technology will be invented to raise health care service staffs work efficiency and performance in any hospitals or clinics in the future.

In conclusion, (AI) technology will raise efficiency for any service or manufacturing industries in the future, although, it is possible that it will also rise low skillful workers unemployment numbers. But, the most important influence to human technological innovation will be risen and it will influence human life will be changed to be better, e.g. self drive cars, health care physician machine men, machine man cleaners etc. intelligent machine men will be manufactured to serve for our daily life. Furthermore, (AI) technological products will influence countries trading, some low technological development countries manufacturing businessmen can choose to buy any (AI) products to raise whose productivity and efficiency and reducing cost to achieve economic cost saving result. Also, GDP of trading growth income will increase to the (AI) products sale countries. Hence, it will be beneficial to economic development to both developed and developing countries both in the future as well as (AI) scientists time and money spending will be valued to continue to invest (AI) technology development for human life and economy benefits for long term.

In conclusion (AI) technology will raise macro economy growth and it can create many (AI) jobs , but it also raise the low level technological worker unemployment change. In the future, (AI) technology can be applied to digital technology to attempt to invent any new undiscovered (AI) and digital technology. So, it needs any scientists to continue to research how digital and (AI) technology can be mixed to satisfy human's future undiscovered needs.

Must Developed And Developing Countries
Different Artificial Intelligent Development Stages

Must developed and developing countries need artificial intelligent development? If one developed country, e.g. US, UK , Japan , Singapore it does not continue to develop artificial intelligence, robotic, then what disadvantges or weaknesses , it will encounter to compare when it chooses to continue to develop this artificial intelligent technology in society. If one developing country, e.g. China, Korea, Taiwan, it does not continue to develop artificial intelligence, robotic, then what disadantages or weaknesses, it will also encounter to compare when it chooses to continue to develop this artificial intelligent technology in in society. I shall explan the reasons why the results may cause to either the developed country, or the developing country as below:

- How AI help developing countries to communication and agriculture and learning and medical delivery development

Why can AI help developing countries ? Drones that pick inaccessible crops and mobile phones that give medical advice are two of the ways AI can transform life in the developing world. Artificial intelligence (AI) may improve the lives of the world's poor, the technology needed to revolutionise inefficient, ineffective food and healthcare systems in developing countries is well. For example, in low-income areas, agriculture and healthcare are two critical ecosystems that we can apply AI to immediately; this is not the far future, or even in five years.

Artificial intelligence (AI) has seeped into the daily lives of people in the developed world. From virtual assistants to recommendation engines, AI is in the news, our homes and offices. There is a lot of potential in terms of AI usage, especially in humanitarian areas. The impact could have a multiplier effect in developing countries, where resources are limited.

Emergency Response to developing countries' earthquake natural damage suddence occurrence predicting

AI and machine learning are still finding importance in emerging markets, but certain applications have emerged and are now widely used. For instance, predictive models for disaster relief enable first responders to automatically analyze large-scale behavior and movement through multiple sources of data including social media platforms, web forums, news sources, etc. Based on collected data, responders can scale reconstruction efforts and distribute supplies in a timely manner.

Why and how AI can assist farmers to predict when the earthquake occurs suddenly in order to avoid or reduce the natural damage to their agriculture productive number loss. For example, In 2015, when a major earthquake hit Nepal, more than 8 million people were affected. During the aftermath, drones were used to map and assess the destruction and speed up the rescue mission. The town of Sankhu, situated about 20 kilometers northeast of Kathmandu, was among the highly affected locations. In May 2018, my company Fusemachines and GeoSpatial Systems partnered with Sankhu's city officials to use drones and artificial intelligence in an effort to automatically estimate the reconstruction need. After processing data accumulated from a drone-powered aerial mapping of the region, the team fed this data to advanced machine learning algorithms. Combining drone imagery, digital mapping and machine learning, the team configured region modeling and infrastructure development with higher accuracy. Another organization known as One Concern, a California-based startup, has created a predictive AI program called Seismic Concern to accurately predict seism and is also working on solutions for wildfires, floods and hurricanes.

Smart AI Agriculture

Another application of AI in developing countries is smart agriculture. Farmers monitor crops more effectively and make better predictions on planting, weeding and harvesting using AI tools. It can also be used to analyze one plant at a time and add pesticides only to infected plants and trees instead of spraying pesticides across large swaths of crops. One California-based tech company is an example of this use of AI. So, the developing countries farmers in rural parts of India are also using AI to increase yields through better access to information about the farming season than they would normally have. Technology-enabled process automation offers the agribusiness industry the chance for remarkable growth -- not only in developed countries but around the world. There's a unique opportunity to increase yields, cut down labor costs and improve people's health.

Medicine Delivery to developing countries' patients urgent need

Companies are also leveraging AI to improve access to health care in some of the most remote areas of the world. In Rwanda, for example, Zipline is using drones to deliver medical supplies and blood to hospitals and clinics that are difficult to access by car. This has dramatically impacted people living in remote parts of the country because they are able to get medical help when needed. The drone system in Rwanda has also helped reduce waste of blood by 95%, as noted by Zipline. One Concern has created an AI program called Seismic Concern that accurately predicts seismic events and is also working on solutions for floods, wildfires and hurricanes. The medical field may actually benefit the most from emerging technologies in developing countries.

Assistance to reduce teaching work workload or psychological pressure to teachers in developing countries' schools

Another vital area benefiting from innovative technologies like AI is education. Advanced technologies can enhance how we learn, teach and perform tasks. In most developing countries, schools lack experienced teachers and resources to enhance students' knowledge. As a result, many students still have to walk long distances to get to the nearest school, which has created education gaps, especially in rural areas. AI tools such as personalized learning assistants can simplify learning by making tutoring services and learning materials accessible to all students, wherever they are. Machines can be automated to help students learn basic concepts without a tutor, which companies like Carnegie Learning are working on. This would allow students to learn at any time from anywhere. With AI, education is made easy and accessible to more people.

The initial usage of AI in developing countries has been at a micro level -- solving small, specific problems in a defined industry. As machine learning advances and there is a higher utilization of AI, we will see more complex issues being targeted and resolved. When duly adopted, AI can positively impact future developing countries people everyday lives not just in disaster intervention, education, health care and agriculture but can also help in mitigating poverty, malnutrition and pollution. Especially, in developing nations, to leverage AI's true potential and create a snowball effect. Startups are defining a holistic and humanitarian approach to building more sophisticated, AI-ready societies. Stakeholders in the AI landscape should understand the strengths and nuances of the developing world as well as the limitations of AI and create localized solutions and applications.

Why does smart phone help developing countries communication ?

Internet Seen as Positive Influence on Education but Negative on Morality in Emerging and Developing Nations. Internet access differs substantially across the 32 emerging and developing countries polled, with the lowest rates of internet use in South Asian and sub-Saharan African nations. Within countries, computer owners, young people, the well-educated, the wealthy and those with English language ability are much more likely to access the internet than their counterparts. To access the internet, people increasingly use smartphones rather than more cumbersome fixed landline connections and computers. Around the world, both smartphones and basic-feature phones alike are used for sending messages and taking pictures.

In fact, many developing countries young people, students are popular to use smart phones for internet usage aim, instead of communication. Moreover, many developing countries working people are also popular to use smart phones for any working usage in their working time , even non working time any time. So, smart phones (AI) phones will be important communication or leisure tools to developing countries people in the future. Unless, it is one day, scientists can develop another new communication tool to replace smart phones. So, artificial intelligence will be important to influence developing countries people , how to improve or bring positive learning attitudes to students in their daily learnnng lifes. as well as how to raise developing countries people, how to raise working people efficiency or improve performace in their daily working lifes. So, AI may bring positive learning or working attitudes to developing countries working people and students both.

The Positive Impact of Mass Media in Developing Countries

Radio, newspapers, television, Internet, social media, etc., all of these are forms of mass media. Each of these outlets has the capability of bringing information to thousands of people with one device. While in some communities it is easy to take advantage of these communication outlets such as television and Internet access, not everyone has access to such outlets. Radio is one of the most common forms of mass media in developing countries because it's affordable and uses less electricity than many other forms of mass media, but only approximately 75 percent of people in developing countries have access to a radio, and roughly 77 percent of people in rural areas have access to electricity.

For developing countries that have implemented forms of mass media in their communities, there have been numerous positive outcomes are influenced to impact developing countries mass media by artificial intelligence as below:

When AI is participated to developing countries mass media, it can influence any radio, television audiences raise more attention to each other through social media platforms such as Facebook and Twitter and create, organize and initiate street protests and campaigns. Furthermore, having access to social media in developing countries, people are able to connect to those that they usually wouldn't have the chance to talk to. Moreover, AI Provides educational opportunities- In many countries, the division between local and national languages as well as issues of literacy can make communication difficult. With the use of mass media, a bridge can be built between these two gaps. In India, there is a radio station that provides information in local languages and respects local culture and traditions. One of the main ways is to create public awareness of what is going on with businesses and government officials. The media plays an important role in giving people the opportunity to act against injustice, oppression and misdeeds that they otherwise wouldn't know about. Information on available healthcare, a mass radio broadcast was sent out encouraging parents to seek treatment at local healthcare facilities for their sick children. With this mass outreach on healthcare, the encouragement of people to take their children to healthcare facilities saved thousands of lives. This easy way of encouraging others and bringing awareness about certain diseases was made possible through a simple radio broadcast. Finally, when AI is particiapted to media, it may bring many social issues to life that otherwise would remain unknown to many people. In developing countries and communities like Burkina Faso, when the radio broadcast was released about malaria, diarrhea and pneumonia, people were educated and moved to action and knew to take their children to healthcare facilities for preventative care. As it is seen, having access to different media outlets is vital for those in developing countries. Here are three ways that those in developing countries can implement mass media to help their people and communities.

When AI is participated to any internet radio or internet newspaper mass online listening or reading channel. It can provide online radios or newspapers in public places- By providing online radios and newspapers in public areas it gives community members to access news, information and emergency warnings. Even though radios can be on the cheaper side, there are still many people that can't afford to have a radio in their home. By providing one in a local place, not only would it better educate the community members but also it will bring the community together. So, it can make media outlets a two-way platform- Creating a two-way platform between the community and those who are behind the radio stations, newspapers or broadcasts makes the community feel involved and that their voices are being heard. An organization called Soul City in sub-Saharan Africa is showing how well two-way platforms work by engaging their listeners and having them contribute thoughts and ideas about complex issues. Because developing countries radio listening audiences or newspaper readers are popular to accept computer online radio listening channel or online newspaper reading channel to replace traditional paper newspapers or radio machines. So, AI may raise their listening news or reading news leisure feeling from online mass media channel in the future.

● Why do developed countries need to develop AI

Artificial intelligence, or AI, is driving massive shifts across the globe, and every day more questions arise. What impact will AI have on the workforce and how can we prepare for it? How can we encourage economy-boosting and job-creating technologies? How can we ensure that AI will be implemented ethically and with minimal bias? How will society benefit? For developed country, such as US example. None of the US, Israel and Russia have a formal national AI policy yet. Private sector companies such as Google, Amazon and Apple and the US department of defence are driving the bulk of AI investment in the United States. Though Israel does not have a specific policy, it is keenly focused on AI and has seen the number of AI start-ups triple since 2014.

Developed country may learn whether what weakness it is lacking when it does not continue to develop AI from one another developed country. Which countries are approaching AI most effectively, and to what degree is there opportunity for greater international collaboration? It may be too early to tell; however, when analyzing the best practices of existing national AI policies, there is much that can be learned. These are the specific areas to consider. When one developed country continue to develop or research AI, it may bring these benefits as below:

On gathering Data aspect, from self-driving vehicles to smart cities, data is the driver behind AI. Innovation in the United States is limited without a national strategy that answers questions about protocol and ownership. France and Denmark, on the other hand, are opening government data. France is hosting troves of centrally collected public and private data that it plans to make available as part of its strategy. Conversely, by taking a restrictive position on issues of data collection (as indicated by the implementation of General Data Protection Regulation), the EU is putting manufacturers and software designers at a disadvantage while balancing the demand for privacy. On raising technologica talent aspect, the demand for AI talent far outweighs the available supply. As a result, almost every nation's strategy addresses talent development. Canada's AI strategy is distinct in that it primarily focuses on research and talent strategy. The country boasts AI degree programmes and is building a $127 million research facility in Toronto. Companies like Facebook and my own company, Uptake, are investing in Canada to access this talent pool. On AI legal technological innovation aspect, a whole host of legal questions swirl around AI. The country is developing a bill for AI liability that will be ready in March 2019. The government hopes the legal framework will attract investors by providing a simple, comprehensive guideline to enable the broad use of AI systems. So, when the developed country applied AI technology to assist any lawyers to work, then AI can help them to reduce the workload to draft any legal documents more easier. So, any developed countries lawyers' draft legal documents time must reduce if the developed countries lawyers accept to apply AI to assist their legal works. One of the great promises of AI is its potential for improving quality of life. But without the right planning and oversight, we risk exacerbating problems of inequality or marginalizing groups of people. As an example, India's AI strategy is focused on leveraging the technology not only for economic growth, but also for social inclusion.

AI may bring what benefits to developed countries

From SIRI to self-driving cars, artificial intelligence (AI) is progressing rapidly. While science fiction often portrays AI as robots with human-like characteristics, AI can encompass anything from Google's search algorithms to IBM's Watson to autonomous weapons. Artificial intelligence today is properly known as narrow AI (or weak AI), in that it is designed to perform a narrow task (e.g. only facial recognition or only internet searches or only driving a car). However, the long-term goal of many researchers is to create general AI (AGI or strong AI). While narrow AI may outperform humans at whatever its specific task is, like playing chess or solving equations, AGI would outperform humans at nearly every cognitive task.

Why research AI safety? Would AI bring war when AI is continued to develop by developed countries? In the near term, the goal of keeping AI's impact on society beneficial motivates research in many areas, from economics and law to technical topics such as verification, validity, security and control. Whereas it may be little more than a minor nuisance if your laptop crashes or gets hacked, it becomes all the more important that an AI system does what you want it to do if it controls your car, your airplane, your pacemaker, your automated trading system or your power grid. Another short-term challenge is preventing a devastating arms race in lethal autonomous weapons.

In the long term, an important question is what will happen if the quest for strong AI succeeds and an AI system becomes better than humans at all cognitive tasks. As pointed out by I.J. Good in 1965, designing smarter AI systems is itself a cognitive task. Such a system could potentially undergo recursive self-improvement, triggering an intelligence explosion leaving human intellect far behind. By inventing revolutionary new technologies, such a superintelligence might help us eradicate war, disease, and poverty, and so the creation of strong AI might be the biggest event in human history. Some experts have expressed concern, though, that it might also be the last, unless we learn to align the goals of the AI with ours before it becomes superintelligent.

There are some who question whether strong AI will ever be achieved, and others who insist that the creation of superintelligent AI is guaranteed to be beneficial. At FLI we recognize both of these possibilities, but also recognize the potential for an artificial intelligence system to intentionally or unintentionally cause great harm. We believe research today will help us better prepare for and prevent such potentially negative consequences in the future, thus enjoying the benefits of AI while avoiding pitfalls.

How can AI be dangerous when developed countries continue to develop AI to become weapon to replace soldiers?

Most researchers agree that a superintelligent AI is unlikely to exhibit human emotions like love or hate, and that there is no reason to expect AI to become intentionally benevolent or malevolent. Instead, when considering how AI might become a risk, experts think two scenarios most likely:

The AI is programmed to do something devastating: Autonomous weapons are artificial intelligence systems that are programmed to kill. In the hands of the wrong person, these weapons could easily cause mass casualties. Moreover, an AI arms race could inadvertently lead to an AI war that also results in mass casualties. To avoid being thwarted by the enemy, these weapons would be designed to be extremely difficult to simply “turn off,” so humans could plausibly lose control of such a situation. This risk is one that’s present even with narrow AI, but grows as levels of AI intelligence and autonomy increase.

The AI is programmed to do something beneficial, but it develops a destructive method for achieving its goal: This can happen whenever we fail to fully align the AI's goals with ours, which is strikingly difficult. If you ask an obedient intelligent car to take you to the airport as fast as possible, it might get you there chased by helicopters and covered in vomit, doing not what you wanted but literally what you asked for. If a superintelligent system is tasked with a ambitious geoengineering project, it might wreak havoc with our ecosystem as a side effect, and view human attempts to stop it as a threat to be met. So, a super-intelligent AI will be extremely good at accomplishing its goals, and if those goals aren't aligned with ours, we have a problem. You're probably not an evil ant-hater who steps on ants out of malice, but if you're in charge of a hydroelectric green energy project and there's an anthill in the region to be flooded, too bad for the ants. A key goal of AI safety research is to never place humanity in the position of those ants.

Why the recent interest in AI safety ?

Stephen Hawking, Elon Musk, Steve Wozniak, Bill Gates, and many other big names in science and technology have recently expressed concern in the media and via open letters about the risks posed by AI, joined by many leading AI researchers. The idea that the quest for strong AI would ultimately succeed was long thought of as science fiction, centuries or more away. However, thanks to recent breakthroughs, many AI milestones, which experts viewed as decades away merely five years ago, have now been reached, making many experts take seriously the possibility of superintelligence in our lifetime. While some experts still guess that human-level AI is centuries away, most AI researches at the 2015 Puerto Rico Conference guessed that it would happen before 2060. Since it may take decades to complete the required safety research, it is prudent to start it now.

Because AI has the potential to become more intelligent than any human, we have no surprise way of predicting how it will behave. We can't use past technological developments as much of a basis because we've never created anything that has the ability to, wittingly or unwittingly, outsmart us. The best example of what we could face may be our own evolution. People now control the planet, not because we're the strongest, fastest or biggest, but because we're the smartest. If we're no longer the smartest, are we assured to remain in control?

A captivating conversation is taking place about the future of artificial intelligence and what it will/should mean for humanity. There are fascinating controversies where the world's leading experts disagree, such as: AI's future impact on the job market; if/when human-level AI will be developed; whether this will lead to an intelligence explosion; and whether this is something we should welcome or fear. But there are also many examples of of boring pseudo-controversies caused by people misunderstanding and talking past each other. When one developed country continue to develop AI, can itself country's all factories workers will lose jobs, due to AI can replace them to do simple works in factories, or any public transport drivers, e.g. bus drivers, ferry , tram, train drivers, they will lose jobs, when AI ( non manual driving drivers ) can replace all public transport drivers. So, some occupations will lose if developed countries continue to develop or research AI to replace human to do some simple jobs, such as some cooking jobs can be done by AI. So, it is possible that future cookers won't be needed, because AI cooking skills may be better than them to cook any good taste chinese or western food in restaurants. If you drive down the road, you have a subjective experience of colors, sounds, etc. But does a self-driving car have a subjective experience? Does it feel like anything at all to be a self-driving car? Although this mystery of consciousness is interesting in its own right, it's irrelevant to AI risk. If you get struck by a driverless car, it makes no difference to you whether it subjectively feels conscious. In the same way, what will affect us humans is what superintelligent AI does, not how it subjectively feels.

In fact, AI may be make any brokers jobs in financial market. the main concern of the beneficial-AI movement isn't with robots but with intelligence itself: specifically, intelligence whose goals are misaligned with ours. To cause us trouble, such misaligned superhuman intelligence needs no robotic body, merely an internet connection – this may enable outsmarting financial markets, out-inventing human researchers, out-manipulating human leaders, and developing weapons we cannot even understand. Even if building robots were physically impossible, a super-intelligent and super-wealthy AI could easily pay or manipulate many humans to unwittingly do its bidding. So, future brokers will be replaced by AI, when AI can be made to own financial brokers' analytical mind to make more accurate whether the share price will rise up or fall down to compare human financial brokers' analytical mind. The robot misconception is related to the myth that machines can't control humans. Intelligence enables control: humans control tigers not because we are stronger, but because we are smarter. This means that if we cede our position as smartest on our planet, it's possible that we might also cede control.

Not wasting time on the above-mentioned misconceptions lets us focus on true and interesting controversies where even the experts disagree. What sort of future do you want? Should we develop lethal autonomous weapons? What would you like to happen with job automation? What career advice would you give today's kids? Do you prefer new jobs replacing the old ones, or a jobless society where everyone enjoys a life of leisure and machine-produced wealth? Further down the road, would you like us to create superintelligent life and spread it through our cosmos? Will we control intelligent machines or will they control us? Will intelligent machines replace us, coexist with us, or merge with us? What will it mean to be human in the age of artificial intelligence?

Why do developed countries people need AI ?

Why do we assume that AI will require more and more physical space and more power when human intelligence continuously manages to miniaturize and reduce power consumption of its devices. How low the power needs and how small will the machines be by the time quantum computing becomes reality? Why do we assume that AI will exist as independent machines? If so, and the AI is able to improve its Intelligence by reprogramming itself, will machines driven by slower processors feel threatened, not by mere stupid humans, but by machines with faster processors? What would drive machines to reproduce themselves when there is no biological incentive, pressure or need to do so?

Who says superior AI will need or want to have a physical existence when an immaterial AI could evolve and preserve itself better from external dangers. What will happen if AI developed by competing ideologies, liberalism vs communism, reach maturity at the same time, will they fight for hegemony by trying to destroy each other physically and/or virtually. If AI is programmed to believe in God, and competing AI emerges programmed by muslims, christians or jews, how are the different AI's going to make sense of the different religious beliefs, are we going to have AI religious wars? What if the "powers that be" greatest fear is the emergence of a super AI that police's and rationalizes the distribution of wealth and food. A friendly super AI that is programmed to help humanity by, enforcing the declaration of Human Rights (the US is the only industrialized country that to this day has not signed this declaration) ending corruption and racism and protecting the environment.Most benefits of civilization stem from intelligence, so how can we enhance these benefits with artificial intelligence without being replaced on the job market and perhaps altogether?

Key to the process of machine learning are neural networks. These are brain-inspired networks of interconnected layers of algorithms, called neurons, that feed data into each other, and which can be trained to carry out specific tasks by modifying the importance attributed to input data as it passes between the layers. During training of these neural networks, the weights attached to different inputs will continue to be varied until the output from the neural network is very close to what is desired, at which point the network will have 'learned' how to carry out a particular task. A subset of machine learning is deep learning, where neural networks are expanded into sprawling networks with a huge number of layers that are trained using massive amounts of data. It is these deep neural networks that have fuelled the current leap forward in the ability of computers to carry out task like speech recognition and computer vision.

In conclusion, when developed countries continue to develop AI, it may bring positive advantages to bring raising productivies, or efficiencies, but it may also raise unemployment ratio to any low skill or low knowledge jobs in ther societies. However, human future society will need to change to be better to raise our living standard. But AI is one kind the best choice tool to achieve this aim in our future, so I agree developed countries continue to develop or research AI to be the super -human machine.

Reference

PWC, Sizing the prize, see: https:// www. pwc. com/gx/en/issues/data-and- analytics/publications/artificial- intelligence- study.html.

Rotman M. J., Data mining-a practical approach to database marketing (1995). IBM.

# FIVE

# THE RELATIONSHIP BETWEEN SOCIAL CHANGE AND HUMAN BEHAVIOR

Human Behavioral network job brings social economic benefits

What does human network job mean ? Why may human network job be popular? Why human network job behavior may influence economy ? Nowadays internet is popular to use. We can apply internet to find data , search any new things, even earn money. Why does internet may become huma network job source. For example, e-publish may be one kind of new human network job. Any authors may apply internet channel to help them to sell electronic or paper books from e-publisher web store. They may apply facebook, you tub etc. any online channel to promote themselves new books to let new readers to know whether when they may buy themselves favourable new topic books to read from electronic publisher web store.

Thus, future electronic publisher industry may help any authors to build internet network platform to help them to sell and promote ot advertise their any one new electronic or paper book topic to let global any one reader to choose to buy their any new topic books from electronic publisher web store easily and conveniently. However, it implies that electronic network platform author may be one kind of future new human network job in our societies.

How electronic network platform author job may bring economy benefit in macro economy view? A person can have few friends, contacts and still be very influential if these few
friends and contacts are themselves highly influential, e.g. one author must not need to know any one reader in global society. When they like to choose any electronic books from electronic internet network platform. They may become the author's any one topic book buyer, when they feel the author's any one topic book is fun and attract they make decision to buth the strange author whose the topic book from electronic book publisher's platform web store conventiently in short time. Although, they are strangers, they do not know themselves , but the reader can understand what it way that made Google from writing platofrm to create new creative mind and typing network job method to replace traditional hand writing book method for global authors. It will be one kind of new human network writing job.

Hence, global any one reader can apply an innovative search engine , such as google.com to find whether whom author personal new topic books are value to read from internet.
Then, the electroniuc publisher's web store may be new book store platform sale network to help the author to sell many electronic or paper books from electronic network platform
in short time. So, internet may be future new network plaform to help global any one author to create network writing job absolutely. Furthermore, internet may be popular social media
to help any one author to build goold relationship between his/her readers. It is one kind of new network, human network job. New authors do not need to buy many paper books to prepare to put in any one book shop warehouse. Their every book can print on demand to reduce out of book stock in any one book shop. They may choose to sell either electronic books or paper books both from any one book publisher web store. So, electronic network platform may be one kind of good writing channel to help human authors to create income and it can also help authors to bring new creative mind and new topic fun content books to let readers to know and buy to read from electronic publisher network platform.

Why does human behavior may be one kind of new human network job to bring global economic advantages. ALthough, it may be free income or without inocme, but the person does the network behavior, his/her behavior may be bring advantages to influence many other people's health. For this case, when a worker in a coffee shop in an airport gets a vaccination aganinst the flu, it does not only helps him or her stay healthy, but also helps the many travellers who might otherwise have been inflected if that workers caught the flu. So, the externality , the result implies the vaccination of even a part of a community conveys benefits to the whole community. For example, governments pay special attention to the vaccinations of school children, teachers, health mothers, and the elderly, categories of people particularly susceptible not only to catching, but also to transmitting a disease.

It is not accidential that governments are heavily involved with vaccination . When there are externalities, free market, fail to persuade individual incentives with society's

their the worker's decision of whether to get a vaccine ends up attracting whether other people get sick. The workers might not fully take all these other people's potential suffering into account when making her or his vaccination decision.

As Stanford University does many suggestions, understand this and tries to help them make the right decisions and so providers free flu vaccines for its staff and students.

Small pockets of unvaccinated individuals can allow a disease to gain a spread more widely well-being. For example, parent weighing the costs and benefits of a vaccine for their child is not always thinking of the consequences of that vaccination to other people. THese are markets in which subsidizing or regulating behavior can make everyone better off. Because the reason for requiring that a child be vaccinated before enrolling in school is not just to protect that child, because each child's vaccination affects others via potential contagions.

Robots take our jobs behavioral and economy influences

Robot job behavior brings economy influences

If one day robots can replace human to do simple, even complex jobs. They will bring what influences to our global societial economy.The popular economic refrain declares that the

global middle class is dying and robots will soon take our jobs, e.g. shopping center customer service jobs, library service jobs, cinema ticket sale jobs, restaurant kitchen cooker jobs,

even, bus drivers, taxi drivers etc. public transport driving jobs, accountant, doctors etc. professional jobs. Whether it is beautiful or petty matter if our future societies have many human jobs can be replaced to do from robots. Businessman must may reduce to employ employees and reduce to pay salary or wage, when robots can be replaced to do their employees tasks. But, societies must bring unemployement rate rises , due to societies will have many people loss jobs when their employers choose to buy robots to serve their clients or do any office tasks or customer service or cleaning etc. tasks.

In micro economy view, employers may save money in long term, but in macro economy view, it will cause unemployment ratio rises , even crime rate rises when there are many people lose

jobs in societies. These models of doom, though, fail to account for the hundreds of businesses riding the waves of change in their industries when robots may be invented to replace human to do many simple , even complex tasks in our future societies.

WE may image that one small factory needs to manufacture fishes canes to sell to supermarket, the small , cheaper stuff and higher margin parts of the fishes manufacture industry. Before, this factory needs to employe many human factory workers need to help every fresh customer makeing the perfect fishing gear, designed for performance, durability, and cost in order to achieve to manufacture every fish cane in whole fished processing manufacturing stages. Every worker needs to spend about 15 to twenty minutes to finish every fish cane , till to delivery to any supermarket to sell. If this fish canes manufacturing factory can apply manufacturing robots to help them to finish any one working tasks , every robot can only spend five minutes to finish whole fresh fish cane manufacturing process. Thus, every robot can

help this factory save 10 to 15 minutes time to finsh every fish cane manufacturing process. IN fact, time is money, because when every robot can help this factory to reduce 10 to 15 minutes time to compare human worker. Then, this factory can finish about 20 fish canes in one hour if it can use robot to help it to manufacture fish canes. Otherwise, if this factory still use human workers to help it to manufacture fish canes, then it can finsh about 3 to 4 fish canes in one hour. SO, the manufacturing efficiency ensures that robots must help this fish manufacturing factory to raise fish canes number more than human workers. So, in robotic behavioral economy view, manufacturing robots must help this fish canes manufacturing factory to raise fish canes manufacturing number and deliver increasing number to supermarkets to prepare to sell every day. Robots can help this fish canes

manufacturing factory bring manufacturing time saving, rising manufacturing efficiency, improving performance and reducing wages expenditure long time advantages in micro economy view. However, manufacturing robots can also bring disadvanages to society, e.g. increasing unemployment ratio, increasing crime rate,
this factory workers will lose jobs and income, they need earn social welfare from government and increasing government finance pressure in short time, even long time in macro economic view.

Stanford University graduate program in economics, Scott lecturer explained that "in demand and supply economic theory for robots supply and demand case, robots supply number increasing may influence human workers demand number decrease. It sometimes calls " the efficient frontier".
No specific human beings were mentioned in any of economics classes. As robots supply and demand in market case, They ( robots) may be purely theoretical " agents" who reached to the most reasonable sale prices in order to persuade any one businessman buyer to make manufacturing robot buying decision whether robots can help him / her to bring how much saving time , saving money, saving cost, improving performance, efficiency economic benefit before he/she plans to reduce workers number when he/ she decides to apply robots to replace human workers in his/her factory or office or any service department, e.g. cinema ticket sale service, shopping center customer service, shopping center cleaning , supermarket customer service etc. service or sale tasks. When robots can replace human to do any one of these tasks in any organizations. So, robots may be human worker agents who reached to prices the way robots would react to a software
command. There was nothing that explained why some people thrived and others did n't or why truly brilliant, hardworking people could fail when much lazier folks succeeded." Having been admitted to the Stanford University graduate program in economics, Scott lecturer hoped to get his answers there.

How robots influence our future social changing? Using the right technology can be a boon to your business in this economy. For internet example, it is easier than ever to find well-matched customers all around the world, to stay in contact with them, and to more quickly design the products they want. If you focus solely on being cutting -edge, though you risk letting the technology
take over what should be very robust relationships with your customers

, employees, and colleagues. IN nowaddays society, technoligical advances and cutomation, personal
relationships in business are more crucial than ever. I mean that robots can not replace human to serve clients to let them to feel more comfortable and passion more easily. For shoe shop case example, if the shoe shop apply one robot to serve its clients to replace human shoe salesperson to serve its shoe customers. Robots ensure that they can not persuade every shoe potential buyer to make shoe buying decision more easily when robots need to contact every shoe potential buyer. The reason is simple, because robots can not touch any one shoe buyer individual emotion very easier.
If the shoe buyer needs the robots to help him/her to choose any right shoe styles when he/she can not feel himself / herself can make the most right shoe style choice decision. The robots can not replace human shoe salesperson to make shoe style choice judgement more easily. They must need longer time to analyze whether which shoe style may be the most suitable to the shoe buyer. Otherwise, human shoe salesperson may attempt to make the most right shoe style choice decision to help any one shoe buyer to chooce the most right style shoe because he/she owns shoe style sale experience, shoe style knowledge, the most important reason is that they can feel every shoe customer individual emotion to touch whether he/she will feel comfortable or happy when they attempt to help every shoe customer to seek the most right shoe style in every shoe customer whole shoe searching processing. Othwerwise, serving robots are only one machine, they can not touch or feel every shoe customer individual emotion whether he/she feel comfortable or unhappy or happy when they need to contact them in whole shoe searching processing. Hence, I believe that some tasks robots can
not repalce human staff to do very easily. Otherwise, robots may bring disadvanatges to let any one businessman to loss his/her customers, due to robots can not touch every customer
emotion to compare human staff in service tasks more easily. Robots serving customer behaviors may cause money lose and customers number lose to the shop in micro economic view.

Intellectual human economic behaviors

What does intellectual human economic behaviors mean ? I believe that when we choose or decide to do intellectual behaviors, then our societies will be influenced to bring economic growth in consequence.I shall attempt to indicate pollution case to explain how and why eithet our intellectual or foolish behaviors may bring economic growth or recession in consequence as below:

On one hand, for air pollution social case aspect example, if we only consider to buy cars to drive for working aimr or holiday leisure aim. Then, our societies air will be polluted. Our health will be influenced to bad. Our car driving behaviors may cause global environment air pollution serously. In long tiem, global air pollution will bring our bodies health to be bad. Although, ourselves car driving behaviors may bring our driving travelling leisure enjoyment and comfortable feeling in short time, also we so not need to pay public transport fare often, but we need to compensate ourselves health economic intangible loss due to air pollution , when cars number increases, dirty air will cause ouselves health to become bad.

In the result, we will need to pay more medical expenditure when we are old age, due to ourselves bodies will become bad, due to we breathe global dirty air every day, due to ourselves cars pollute air in long time, e.g. 10 to 20 years, even 30 more without limited air pollution environment. So, driving cars behavior may be one kind of human foolish behavior and our foolish behavior may bring ourselves future long time medical expenditure absolutely.

One the other hand, water pollution social aspect, if we often keep much rubblish to pollute sea, oil exploration porcessing pollute ocean , ships gas pollute ocaen, then fishes will eat polluted food and drive dirty water, due to global ocean is polluted.

In fact, because human only to conside how to buy boats to carry on leisure enjoyment activities, or catch cruises to travel on the sea. Also, oil manufacturers only consider researching anywhere to find new oil exploration places to manufacture oil product, when their oil exploration processes pollute ocarn . Consequently, global fishes drink polluted warer or eat polluted food. They will have poison. SO, human will have high chance to eat poison polluted fishes, due to fishes are poison or are polluted.

So, human is doing foolish activities, we only hope to find oil exploration places to pollute ocean or we only spend money to buy ticket to catch ships to travel anywhere in global ocean. All of these human foolish behaviors will bring pollution to global ocean. On consequently, we will need to compensate to eat polluted or dirty or poision fishes, ourselves bodies health will be bad. In long time, we need have high chance to pay medical expenditure when we are old. So, pollution case may be one good example to explain how and why human foolish behavior may influence ourselves future need to compensate serious medical loss.

All of these human foolish behavior will bring pollution to global ocean. On consequently, we will need to compensate to eat polluted or dirty or poison fished , ourselves bodies health will be bad. In long time, we will have high chance to pay medical expenditure, when we are old. So, pollution case may be one good example to explain how and why human ourselves intellectual or foolish behaviors may influence future long time economic loss or economic growth or recession in micro and micro economic view.

On another water pollution aspect hand, if we often keep rubbish to sea, oil exploration processing pollutes ocean and ships' gas pollute ocean, then fishes will eat polluted food and drink dirty water, due to fishes will eat polluted food and drink dirty sea water because the global ocean is polluted seriously.

In fact, because human only consider how to buy boats to carry on any leisure water activities, or catches cruises to travel on the sea. Also, oil manufacturers only consider any where to find oil exploratin places to manufacture oil products from ocean, when their pol exploration processes can plooute ocean. Consequently, global fishes drink polluted water or eat direty food. They will have poison. So, human will have high chance to eat poison fishes.

Otherwise, such as pollutin case, it can infuence inflation or deflation. Consequently, the reason indicates supply and demand theory. If air pollution is serious, then we will consider health issue, global cars demand number may be influenced to reduce, when global cars number demand will reduce, global car prices and supply number will need to change to fall down in order to attract or persuade global car consumers choose to make car purchase decision.

Hence, global car manufacture number and car price will be influenced to reduce, due to global air pollution issue. Consequently, deflation will occur because when the country citizen usually does not spend much extra saving money to buy car expensive goods. Money value will be low. Otherwise, if global cair pollution is not serious, human considers to buy cars to enjoy driving leisure lives. So, global car demand is influenced to increase , also global car price will also influenced to increase.

Consequently, gobal human will choose to buy cars to drive. Due to we accept to spend extra saving to buy expensive car goods. Car sale price and supply may be influenced to rise up. Money value is influenced to reduce. Inflation may be influenced, due to global car consumers number increases, we would not have extra money to spend easily. Car expensive goods expenditure influences our spending habit to avoid to make car purchase decision more easily. So, human intellectual or foolish activities may bring inflation or deflation consequency in possible indirectly in macro economic view.

On conclusion, above pollution case explain that how and why human intellectual or foolish economic behaviors may bring inflation or deflation consequency as wll as economic growth or recession consequency as well as any goods demand and supply increasing or decreasing consequency. It implies that human behavior may have indirect relationship to influence any goods demand and supply number to either increase or decrease result as well as any goods price will be influenced to increase or decrease in micro and macro economic view.

The relationship between social change and human behavior

Why does economic changes may influence human individual behavioral change? I shall attempt to indicate shopping behavior and staying at home behavior to explain their case and effect relationsip as below:

Human behavior can be influenced by economic change or economic change can be influenced by human behavior? Why does recession may influence consumers reduce shopping desire? In social recession suitation, it is possible that many people lose jobs suddenly, due to businessmen lose many customers. They need to make decision to reduce employees number in order to continue to keep businesses. Consequently, many firms ( organizations) their employees may lose jobs. When they have much time, due to lose jobs, they will feel to avoid to spend too much time and money to go to shopping often. Many losing jobs people, they will often stay at homes. So, they will reduce time to go to shopping, then non essential products won't their preferable choice purchase products. Hence, recession will change many losing jobs people their shopping or consumption desires to avoid to buy non essential products often . Usually when economic boom, many people have jobs to do because consumers number must increase when many people have jobs to do. Then, many people can accept to spend money to buy non essential products often. Many people feel spend time to go to shopping can satisfy their purchase of any kinds of new products useful psychology or desire. So, recession is one good example to explain it can influence many people do not like often to leave homes to go to shopping easily. Many people like to stay at homes, becaue they feel worry about spending too much shopping time when they leave homes. Their staying home time is one good negative shopping behavior example. So, economic change may influence human individual behavior changes , they have direct cause and efect relationship in behavioral economic view.

May human behavior influence economic change? Is it possible that human behavior may bring the country social economic change in macro economic or micro behavioral economic view ? I shall indicate publishing industry example. Do you feel that if there are many students feel learning is very important when they read many books or many of students feel interesting to read or they have reading new books in habit, then it is possible that the country will have many students like to spend time to go to any book shops to choose the books, they feel that they can help they learn new knowledge. Then the country will increase students number, they often spend time to visit any one book shop every week. Their visiting book shops behavior which may become their habits. So, the country will increase students number, they often spend time to visit book shops. Also, it implies that visiting book shops behaviors may be their behavioral habits.

So, when the country has many students often spend time to visit book shops , their visiting book shops behaviors may help any one book shop to raise books sale chance. So, the country's student individual often visiting book shop behaviors, their habitual visiting book shops behaviors must may assist help any one book shop to increase books sale number absolutely.

Consequently, any one book shop , its books sale bumber must be influenced to increase to increase because the country will have many students like or feel need visit book shops habit in order to choose any suitable books to buy to read at home in order to raise themselves learning effort. When the country has many bok shops often have many students visit their book shops, then their books sale number may be influenced to increase. It explain why student individual visiting book shop behavior may help any one book shop sale number increases also.

www.ingramcontent.com/pod-product-compliance
Ingram Content Group UK Ltd.
Pitfield, Milton Keynes, MK11 3LW, UK
UKHW041857190726
13854UKWH00002B/958

9 798887 836188